a search for order
in the physical universe

# a search for order in the physical universe

Clifford E. Swartz     Theodore D. Goldfarb

*State University of New York at Stony Brook*

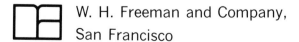

W. H. Freeman and Company,
San Francisco

**Library of Congress Cataloging in Publication Data**

Swartz, Clifford E.
    A search for order in the physical universe.

    1. Force and energy.  I. Goldfarb, Theodore D.,
1935–    joint author.  II. Title.
QC73.S89                    531'.6                    73–19743
ISBN 0-7167-0345-9

Printed in the United States of America

9  8  7  6  5  4  3  2  1

# contents

# preface

This book grew out of a course for nonscience majors that we taught jointly for three years. Although our students were not primarily interested in science or technology, most of them had studied enough mathematics so that they could follow simple algebraic derivations and interpret graphs. They had also been exposed to the standard presentations of one or more high school science courses. It is our conviction that such students are ready for a more challenging experience than would be provided by a wide-wonderful-world survey course.

With such an audience it seemed appropriate to try a different approach to the study of the physical universe. The theme is a study of interactions and a search for conserved quantities, particularly energy. This is the modern approach of scientists working with atomic and nuclear systems; energy becomes a fundamental quantity and force is a derived parameter. Although this method is novel at the introductory level, it is surprisingly powerful in explanation. It is also surprisingly natural and easy for a nonscience major to follow. In addition, our theme is developed with a combination of plausibility arguments and graphs that require no trigonometry and only the simplest of algebra.

After an initial chapter of setting the stage for our investigations, we present phenomena that are most easily understood by defining certain quantities that do not change during interactions. First, kinetic energy and momentum are defined in this manner. In order to extend the range of phenomena governed by such laws, the characteristics of the defined momentum are generalized, and forms of energy other than kinetic are defined. To preserve the energy conservation rule, we then present a model of the microstructure of matter using the study of heat as an approach to atomicity.

Many of the phenomena to be studied can be investigated firsthand by the reader. We have built in such investigations, most of which can be done without any formal laboratory. Since many of our students were preparing to teach in elementary school, we included some demonstrations and experiments that can be successfully done by younger children.

Throughout the text we inserted questions intended to make the reader pause and challenge what the text has just claimed. Our answers to these questions are included at the end of each chapter. Of course, a student can always bypass these questions or turn immediately to our answers. Comprehension seems to be improved, however, if the reader tries first to work out a personal answer for each question, preferably in writing.

Some people fear science because of its aura of cold precision. Actually, scientists usually avoid excessive precision and make frequent use of order-of-magnitude calculations. We frequently appeal to such arguments in the text, and in the appendices have included sections on precision and calculations. In the same spirit, we have tried to avoid fussiness in definitions and classifications. For instance, the words *speed* and *velocity* are used interchangeably, as indeed they are by most research scientists. Similarly, although we make use of vector properties in terms of geometric constructions, we avoid the terminology of vectors.

Wherever possible, the problems and examples given in the text are taken from the real world, and in that sense the material is relevant to everyday life. The theme of the book, however, is not on the technical problems of society or on preparation for employment. Although we agree that such courses can be useful, we believe that attempts to apply science to real problems can be carried out more realistically by students who have been introduced first to the fundamental principles and philosophy that are the foundation of all science and technology. We hope that this text will be of help in meeting this basic need.

The authors wish to thank Professor Mario Iona, Department of Physics, University of Denver, for many valuable suggestions made while reviewing the original manuscript. We also want to thank Mrs. Dorothy Rhame of Stony Brook for typing the text manuscript.

September 1973                                    Clifford E. Swartz
                                                  Theodore D. Goldfarb

a search for order
in the physical universe

# to comprehend the universe

The primitive mysteries remain. We do not know why the universe was formed or how it will end. We do not know whether many of the structural details of the universe and events in its past history were accidental or inevitable. We do not know why there are living creatures, and most important for man, we do not know if there is some purpose or special role for thinking creatures. We do not even know whether it is important that we find out.

Of course, there is a lot that we have found out about the universe in the last few centuries, and especially in the last forty years. It is the business of science to measure the diameters of atoms and the distances to galaxies. It is also the business of science to catalog the objects with sizes in between, and to analyze their interactions. This enterprise has succeeded brilliantly, with experimental discoveries and theoretical revelations tumbling upon us. There is a popular legend that scientific information doubles every seven years or so, and therefore the student of today faces the impossible task of learning all the new facts. We would indeed be drowned by all this information if science did not at the same time provide us with simplifying generalizations. The jumble of recipes and special conditions needed by the alchemist have been systematized by the atomic models of the modern chemists. The complex cycles and epicycles needed by Ptolemy to compute planetary positions have been replaced by the conceptually simpler orbits of Copernicus. It is no more difficult to study science today than it was ten, fifty, or three hundred years ago. How could it be otherwise? Humans remain essentially unchanged, and humans create science.

*Question 1.1*    Throughout the text we will interrupt the line of argument by proposing questions or challenges. At the end of each chapter we have given our own answers with which you may not always agree. If you turn immediately to our answer you might profit by the rest taken from your reading as you hunt for the appropriate page. But, we believe that you will profit a lot more if you try to answer the questions for yourself before seeing what we said.

Do you really believe that new generalizations of science compensate for the increase in the number of discovered facts? Take any particular generalization with which you happen to be familiar: conservation laws, the nature of electricity or light, or genetic laws, for example. List the separate detailed facts that you can now deduce from that generalization, but which could not have been related three centuries ago.

In this crucial role of explanation and simplification, science becomes an art. It is a particularly human art, requiring skilled command of standard techniques and the daring to depart from them. The fact that the scientific enterprise requires such qualities reveals something about science, and also something about the relationship between humans and their universe. The traditions of western civilization have separated humanity from the rest of the universe. Were not humans, according to the first chapter of Genesis, given "dominion over the fish of the sea, and over the fowl of the air, and over the cattle, and over all the earth"? In the Platonic Greek tradition the world exists in an ideal form outside us. We learn about it, step by step, always approaching closer to some final truth. Perhaps this is not the way things really are. Perhaps what we observe and comprehend is determined more by our human nature than by any world external to us.

In this book we will not be primarily concerned with a detailed description of the physical universe. Any particular fact about this subject or even the great generalizations may or may not be relevant to your life. The problem that does concern us is how people go about comprehending their universe. Is it simply a matter of solving a gigantic puzzle, with the clues lying about and one complete answer waiting for us when we have sufficient facts? Or, do we find the clues to mold the generalizations in our own image? The science of this century has exposed the dependence of the measured fact on the measuring technique, and of the theory on the human or even the political cultural nature of the theorizer. A study of how we comprehend our universe is a study of people as much as it is a study of the universe. It is relevant to humans not only because of the technology that springs from science, but because of the view that it presents of the nature of the universe, of human beings and of their interaction.

The study of science is part of a liberal education. If science consisted merely of the technological fruit of the enterprise, it would be hard to claim that knowledge of science is in any sense "liberating." As long as the phones work and the planes fly on schedule, the educated person does not necessarily have to understand the engineering marvels involved. There is more to science, however, than just gears and tubes. The discoveries of science in the last few centuries have completely altered our understanding of ourselves. Particularly in this century these altered views and the powerful methods of scientific analysis have had a profound influence on philosophy. The study of the microstructure of matter erases the ancient dividing lines between matter and energy and raises in a very sharp way the Platonic questions of reality. Analysis of objects at the edge of the universe, and perhaps at the beginning of time, places humans and their works in a new and different perspective. To understand our human complexities, we must also understand some of the complexities of the universe around us.

Paradoxically, we shall try to understand these complexities by looking at very simple things. Simplicity of description and understanding is usually hard to achieve, as we shall see. The motions of the planets around the sun, for instance, are nearly circular—but not as viewed directly by anyone on earth. Their paths, as viewed in our night sky are very complex. It took detailed knowledge of these observations, plus the imagination of genius, to propose the simple system of elliptical orbits. In Chapter 2 we will look for the simplest event that we can find in order to analyze and thoroughly understand it. Before doing this, however, we should take a look at just a few of the raw facts that are known about the universe. The aim of this stage setting is not to teach the facts but to indicate the nature of the problem.

## 1.1   the universe as a subject for investigation

**Scale.**   The range of sizes and distances in our universe is difficult to comprehend. As we investigate the micro-world, we find living cells with diameters 100,000 times smaller than the height of a human being. These cells are made of molecules. In turn, these molecules are made of atoms 100,000 times smaller than the cells. The atom itself has a structure, consisting mostly of "empty" space, electrons, and, in the center, a nucleus 100,000 times smaller than the atom. If we turn to the macro-world, a factor of 100,000 times the height of a human being brings us to a distance of a little over 100 miles, a distance with which we are somewhat intuitively familiar. But another factor of 1,000,000 stretches our imaginations with a distance about equal to that of the earth from the sun. The distance of the nearest star to our solar system is yet greater by another factor of 100,000. Our sun is merely one of some ten billion stars in a spiral galaxy with a diameter 25,000 times that of the distance between the sun and the next nearest star. Over a billion of these galaxies are within range of our telescopes. (See Fig. 1-1.) They stretch out to the very edge of visibility, and perhaps to the edge of the universe, at a distance 100,000 times greater than the diameter of our own galaxy.

figure 1·1   This is the great galaxy seen in the constellation of Andromeda. It is so large and bright that, even though it is two million light years away, it can be seen with the naked eye as a diffuse spot in the sky. (A light year is the distance that light travels in a year; it is equal to $9.4 \times 10^{12}$ km, or $5.87 \times 10^{12}$ miles.) This galaxy is about 100,000 light years in diameter, and contains about ten billion stars. Our own island universe, the Milky Way, is also a spiral galaxy very similar to this one. (*Courtesy of the Hale Observatories.*)

Since we cannot really comprehend such vast differences in size, let us consider scale models of sections of the universe. Imagine a device that could uniformly expand a single raindrop until it is as large as the planet Earth. An atom in the water drop would then be about human size, but its tiny nucleus (which contains almost all of the atom's mass) would still be only about the size of a bacterium—too small to see without the aid of a high-powered microscope! Now reversing our device, let us imagine the shrinking of our galaxy (a very *small* part of the entire universe) until its outer limit would just fit within the space defined by the earth's orbit around the sun. The earth would then have a diameter about as wide as the head of a pin and its human inhabitants would have atomic dimensions!

Figure 1-2 is an attempt to illustrate the range of sizes and distances encountered in the universe. Note that the *unit* of length used is the *meter*. This unit is part of the *metric* system which has the advantage of being a decimal system (based on ten) just like our number system. Other metric units of length are the *kilometer* (1,000 meters) and the *centimeter* ($\frac{1}{100}$ of a meter). The units of mass (*gram*, *kilo*gram, *centi*gram, etc.) in this system are also decimal. Since it is the official system of measurement in much of the world and is used almost universally by scientists (as well as being used extensively in this book) it is worth spending some time familiarizing yourself with metric units. (See the problems at the end of this chapter.)

The scale in Fig. 1-2 is a *logarithmic* or *powers-of-ten* scale. This permits representing an enormous range of magnitudes on a single diagram or graph. We will make frequent use of powers-of-ten (*exponential*) notation since it is much more convenient to write numbers such as 4,320,000,000 as $4.32 \times 10^9$, or 0.00000001576 as $1.576 \times 10^{-8}$. If you are not familiar with this form of notation and with the simple algebraic manipulations (addition, subtraction, multiplication, and division) of numbers written this way, turn to Appendix 1 and practice the exercises given there. Note that on the scale of Fig. 1-2, every step represents an increase or decrease of length by a factor of 10. Two steps means a change by a factor of 100, three steps a factor of 1,000, six steps a factor of one million, and nine steps a factor of one billion. Thus the north-south width of the United States is about one million times the height of a man.

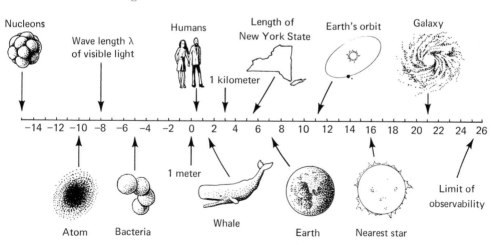

figure 1-2   Map of the Universe. The scale is logarithmic. Each step to the right represents an increase of length by a factor of 10. The unit of length is the meter; hence, 0 (signifying $10^0$) represents 1 meter. Similarly, 2 (signifying $10^2$) represents 100 meters.

*Question 1.2*    Verify the positions of: a human, a whale, the length of New York State, and the size of the earth as shown in Fig. 1-2.

*Question 1.3*    How many times longer would Fig. 1-2 have to be if the scale were linear, rather than logarithmic, and arranged so that $10^{-15}$ meter represents 1 meter?

Look at the place of humans on the map of the universe. We and all the living objects on earth occupy only a small region between the very small and the very large. The largest of all these creatures is less than twenty times as long as a human. The smallest of these objects that still show some of the characteristics of life is smaller than we are by a factor of ten million. Beyond these limits the universe continues to stretch out. Our immediate perception of the universe is severely limited. With the unaided eye we can see objects that are only $\frac{1}{10}$ millimeter across ($10^{-4}$ on our scale). Below that size we can see only with microscopes which extend our view only to objects one hundred to one thousand times smaller. For large objects we have some intuitive comprehension of size up to a factor of $10^5$ larger than ourselves.

To go beyond the normal human grasp of sizes, we must use instruments and interpret their readings in terms of theories that are often complex. The image presented by a microscope is so close to the visual image which we usually see that the only theory needed for comprehension is one that gives us an explanation of magnification. (The concept of magnification cannot be comprehended by most children under the age of ten; the effect is not elementary.) Electron microscopes also produce visual images of a familiar kind, although we may feel uneasy about "seeing" objects with electrons. Of course, the electrons are serving as probes just as the particles of light serve as probes, as shown in Fig. 1-3. In both cases, the probes are transmitted, or scattered, or absorbed. Probes that are even smaller can be used to examine smaller objects, such as atoms and sub-atomic particles. In cases where these small particles are being "seen," the analysis of the scattering and absorption of the probes requires mathematical treatment, and interpretation in terms of theories of particle behavior. However, the pictures that are obtained can be very detailed. Objections are sometimes raised when a scientist claims to have "seen" a subatomic particle by virtue of the visible trail it has left behind in a bubble or spark chamber. The same reaction is not likely to greet the individuals who talk about having seen a movie star when they have actually been observing an image on a movie screen. It is possible to argue about whether a sensory image is caused "directly" or "indirectly" by the subject of investigation but it is certainly preferable to avoid such semantic quibbles. The devices employed in scientific investigations can best be thought of as tools that extend the power of the human senses.

While we might have no intuitive feelings for distances much larger than

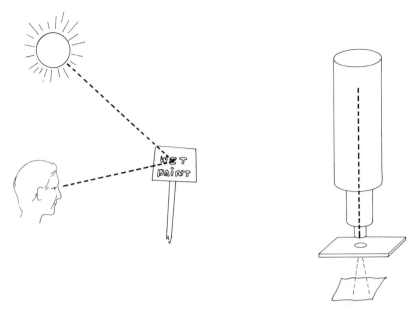

figure 1-3   The "probes" that we use for seeing might be photons (light radiation) that are scattered by an object and are detected by our eyes, or they might be electrons that are scattered by an object and are detected with a photographic film. In either case, when our brain processes the scattering pattern, we have "seen" the object.

the size of a state or of our country, our reach beyond these distances can be extended appreciably with ordinary surveying techniques. These techniques consist essentially of sighting on a distant object from two different points. The length of the line between these points and the sighting angles from each of this baseline's end-points can then be related to the unknown distance. Fig. 1-4 illustrates the geometry involved.

*Question 1.4*    Suppose the surveying base line on the earth is 1,000 miles long. Will an observer at one end of this base line sight a particular point on the moon at a different sighting angle than a person at the other end? Assume the first observer points his telescope directly up; will the second observer's telescope be far from the vertical? The distance to the moon is about 240,000 miles. (Assume that the earth is flat; the correction for the earth's curvature could be made in an actual experiment.) What is the answer to this question if the particular point being sighted were on the sun, 93 million miles away?

In spite of the small angles involved, Greek astronomers at Alexandria, 2,300 years ago, determined the distance to the moon and sun, and their relative

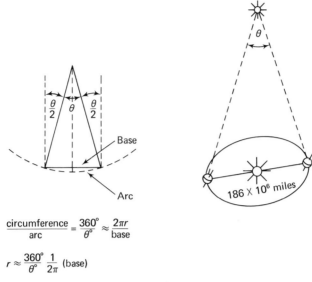

$$\frac{\text{circumference}}{\text{arc}} = \frac{360°}{\theta°} \approx \frac{2\pi r}{\text{base}}$$

$$r \approx \frac{360°}{\theta°} \frac{1}{2\pi} \text{ (base)}$$

figure 1·4    From the baseline, the distant object can be sighted and the angles from the perpendicular direction measured. For a small apex angle, $\theta$, the base is approximately equal to the arc of the circumscribed circle, and the distance to the object is approximately equal to the radius. The diagram scale of the solar system-star geometry is greatly exaggerated. The real $\theta$ for the nearest star is only 0.0005°.

sizes, with surprising results. Although their value for distance between the earth and the sun was wrong by a factor of 60, their estimate of the distance to the moon was wrong by only 25%. Not only did they use angular differences in sighting to get their estimates, but they also used observations of the appearance of the earth's and moon's shadows during eclipses. Notice that they did all this 2,000 years before the invention of optical telescopes!

Surveying techniques can be used to measure the distance to a few of the very close stars. As we saw in Question 1-4, no base line on the earth is long enough to produce a measurable difference in sighting angles to a star. (Compare the distance to the nearest star on the scale map of the universe with the radius of the earth's orbit—which is the distance from earth to sun.)

*Question 1.5*    What base line can be used to measure the distance to the nearest star? About how large is the angular difference in sighting angles when this base line is used?

Very elaborate surveying techniques, using a base line that depends on the motion of our sun among the stars, can extend our distance measurements

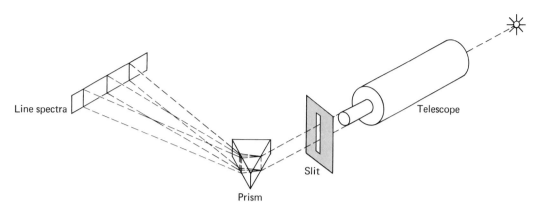

Line spectra

Telescope

Slit

Prism

figure 1-5   Light from a star is focused by a telescope onto a spectroscope. A prism (or a diffraction grating) breaks the light into its component colors. Instead of a complete rainbow of colors, a series of certain lines of different colors appears. (Lines are formed because of the geometry of the slit.) The various patterns of the spectral lines indicate which elements produced the light, and also yield information about the motion of the star.

a relatively short distance into our galaxy. Beyond that, as the map in Fig. 1-2 shows, there are still enormous ranges of size. The methods for measuring the distances to other galaxies depend on theories of atomic processes and on the behavior of light when its source is moving with respect to the observer. Raw data for these measurements is obtained with telescopes, but the star light is not simply observed or recorded on film. No amount of magnification can change the appearance of all but a few of the stars that we see. They are so far away that their light always appears to come from a point. What we learn about them is mostly derived from the nature of the light received —its color and intensity, and fluctuations in that intensity. Telescopes focus light on spectroscopes, instruments which spread out the light according to its color, or wavelength. Most of our information about the stars and galaxies, including our knowledge of their distance from us, is deduced from this information. A simple spectrograph is illustrated schematically in Fig. 1-5.

In other sections of the book we examine in more detail some of the small and large regions of our universe. We also study some of the phenomena that allow us to probe these regions and the theories with which we interpret the experimental results. In this preliminary look we only want to emphasize the vastness of the universe that lies within us and beyond us.

**Complexity.**   In addition to presenting a fantastic range of sizes and distances, the objects in the universe display an innumerable variety of other measurable properties. In an effort to simplify and systematize their investigations scientists often categorize or classify the materials they study. A familiar classification scheme is the attempt to label every substance according to its so-called state-of-matter—gaseous, liquid, or solid. A fourth state called *plasma*, containing high temperature, electrically charged particles, has only recently become accessible for study on earth. This fourth state-of-matter is of great importance elsewhere in the universe. Stars, for instance, consist of plasmas.

Actually, only fuzzy boundaries exist between these states. Considerable attention has recently been focused on *liquid crystals*, a class of liquids that display many of the properties of a crystalline solid. Common window glass and many plastics have both liquid and solid properties. A gas, under conditions of very high pressure, behaves like a liquid in some ways.

Another scheme for the classification of matter is based upon defined distinctions between *pure substances* (elements and compounds), *solutions*, and *mixtures*. Again, fuzzy boundaries are encountered. Certain *solutions*, (e.g. 95% alcohol, 5% water) have sharply defined boiling or freezing points—one of the criteria used to define *pure substances*. Compounds of certain metals with hydrogen have a variable composition and are therefore *pure substances* by certain criteria and *solutions* by others. Colloidal suspensions of very fine particles in a liquid, or of two (or more) immiscible liquids (homogenized milk, for example) are mixtures that have properties that are very similar to those of solutions.

Efforts aimed at describing certain of the properties of an entire class of substances are useful but invariably an individual object within any single category of any classification scheme will present a unique combination of individual properties (e.g., electrical conductivity, chemical reactivity, elasticity, color).

*Question 1.6*    How many other schemes for classifying matter can you think of? Do they also have "fuzzy boundaries"?

**Interactions.**    Most of the observations made on objects in the universe involve the *interaction* of one material with others. We note that the level of water rises in a glass tube with a small diameter, whereas the level of mercury is depressed (see Fig. 1-6). This must involve a difference in the type

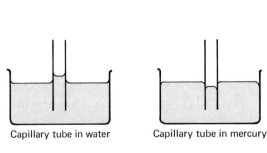

Capillary tube in water        Capillary tube in mercury

figure 1-6    Water is attracted to glass and wets it. At the boundary the water rises onto the glass, making the surface concave as seen from the side. Mercury does not wet glass. At the boundary between mercury and glass, the mercury surface is depressed and convex.

of *interaction* that occurs between glass and water as compared to that between the glass and mercury.

Studies of color involve the investigation of the *interaction* of light with matter.

Other types of *interactions* include: electric and magnetic attractions and repulsions; the effects of mechanical pressure (produced by the weight of an object, the compression of a spring, or the flexing of a muscle); and, the wide variety of atomic and molecular *interactions* called chemical reactions.

## 1.2   attempts to "understand" the universe

**The Scientific Endeavor.**   Confronted by a universe presenting systems for study in vastly differing sizes, with a fantastic variety of properties interacting in all manners of ways, humans try to seek order and achieve some level of understanding out of this seeming chaos. An expensive watch movement (Fig. 1-7) is a complicated mechanism when viewed as a whole. But when the parts from which it has been constructed are spread out, and examined individually, each one is found to be quite simple. The problem, of course, is to discover how they fit together—how they interact! Thus, the scientist carefully observes the basic components and interactions involved in the complicated systems or events being investigated. Simplification is achieved by constructing a model or theory that will explain something complex in terms of interactions among things that are more simple. The theory, once formulated, will inevitably permit predictions concerning the behavior of the same (or a similar) system under a somewhat modified set of circumstances. These predictions must be tested. If they are found to be valid the usefulness of the theory is increased—if not, the theory must be abandoned, or modified to permit an explanation of the new observations in addition to those on which it was originally based.

The explanations of nature that result from scientific activity are based on the observation of behavior that involves *measurable* quantities. The level of understanding achieved in this way has been termed an *operational* one. It forms the basis for an entire philosophy called *operationalism*, in which all concepts derive their meaning from the results of measurements performed on the things to which they refer. An operational definition of anything is given in terms of the behavior of the object, or of what happens when you do something to it. Suppose, for instance, that you had to explain the meaning of "energy." It would be meaningless to define energy as "the ability to do work." What is work, and how do you measure "ability?" Actually, it is difficult to define energy in any short, simple way. To be operational, the definition would have to spell out how the quantity called energy could be measured, and how different methods of measurement have to be used in different circumstances.

figure 1-7   Each part of a watch is a relatively simple device. The parts scarcely hint at the sophisticated instrument that they can become when properly organized.

*Question 1.7*   Give an operational definition of (a) a needle, (b) a martini.

**The Search for Fundamental Quantities.**   The ultimate degree of simplification and understanding would be a complete description of nature in terms of a limited number of *fundamental* particles and the types of *fundamental* interactions they can undergo. All substances found in nature are now known to be the result of chemical combinations of only about 90 naturally occurring elements. These elements have particles called atoms as their basic units. In turn atoms are known to be the result of interactions among even more fundamental particles such as electrons, protons, neutrons, mesons, among others. Fig. 1-8 illustrates some of these relationships.

At present, there are only *four* fundamentally different interactions known to exist in the universe (Fig. 1-9):

1   Gravitational—not very significant for interactions between small particles, but responsible for holding solar systems and galaxies together.
2   Electromagnetic—primarily responsible for the structure and behavior of the atom (including its chemical properties), and also the source of the electromagnetic spectrum. This consists of visible light, as well as invisible forms of radiation

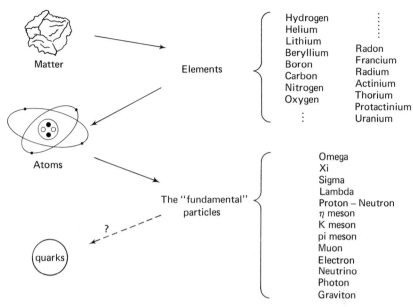

figure 1·8   Matter can be subdivided into ever more fundamental parts.

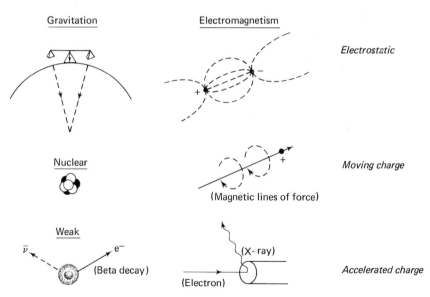

figure 1·9   So far as we now know, there are only four types of interaction in the universe. Note that the electromagnetic phenomena can be classified into three realms: electrostatic, where the electric charges are not moving; electric current, where the charges move with constant velocity and produce magnetic fields; accelerated charge, where the charges radiate away energy in the form of light, radio waves, or X-rays.

(x-rays, infrared and ultraviolet radiation, radio waves, etc.). Electric and magnetic effects are part of the same phenomenon, so they are linked together by the term "electromagnetism."

3   Strong nuclear—holds the atomic nucleus together and also provides the energy to keep the sun, stars, and nuclear reactors going.

4   Weak nuclear—governs the processes by which certain of the subatomic particles are transformed into others.

How can there be only these four interactions? What about air pressure, glue, muscles, and springs? It turns out that all of these can be explained in terms of the electromagnetic interaction. After all, if electromagnetism holds the atom together, and also is responsible for all the chemical effects such as molecular binding, then it must also be responsible for complex molecules; including living cells. Indeed, in our daily life we experience only gravitation and electromagnetism. The other two interactions make themselves apparent only in nuclear research. Gravitation and electromagnetism are long-range forces. Their influence can extend throughout the universe. The strong and weak nuclear interactions take place only within distances of the order of nuclear dimensions.

*Question 1.8*    How is it possible that the gravitational interaction can be insignificant compared to other interactions between the fundamental particles when they are very close to one another, and yet strong enough to be responsible for interactions between stars and planets over vast distances?

## 1.3   our present state of ignorance

Although much is known about the infinitesimal building blocks of the universe, the four basic interactions that they can undergo, and some of the basic rules that control their behavior; science is by no means on the verge of a complete description of nature. Even on the subatomic scale new fundamental particles continue to be discovered and many mysteries remain concerning their detailed behavior.

Turning to a vastly more complicated problem, such as the human organism, one finds, as might be expected, a much less detailed understanding. Considerable knowledge has been gained about the overall functions of the organs and tissues and how they interact. Attempts to describe the functions of these organs in terms of their individual living cells is a task that is still far from complete. The still more basic problem of examining the workings of the individual living cell on the basis of its myriad chemical constituents fashioned from $10^{14}$ or so atoms, is receiving much attention from biochemists and biophysicists. Although progress is being made, they will certainly be at it for a long time to come.

On a cosmic level an unbelievable amount of detailed knowledge has been gained about the constitution, motions, and life-cycles of stars, and other celestial objects. Humans have personally explored the surface of the moon and have sent exploratory probes to nearby planets. However, we are still quite uncertain about the existence of life outside our solar system, the origin and shape of the universe, and even about the composition and structure of the interior of the earth.

## the aims and organization of this book

The choice of topics dealt with in this book, and the viewpoint toward them, is different from that of a standard physics, chemistry, or earth science text. Having presented a brief picture of the vast size and complexity of the universe, we will study, in the succeeding chapters, the simplest phenomena we can find. The interactions involved in these phenomena are governed by some general rules which have surprisingly broad applications to much larger and more complicated systems. One of these rules concerns energy, and in the process of accounting for the energy apparently lost in most interactions it will be necessary to study heat. This search will lead us into a study of the microstructure of matter. The whole universe will then be viewed as a collection of a few types of elementary particles, interacting in only four ways, and subject to only a small number of fundamental laws.

*Answers to Questions*   **1.1**   Consider what we now know about the origin and nature of lightning. Neutral droplets of water can become electrically charged. In certain clouds rapid air movement can concentrate these charges into large, but isolated, regions. A region of moist air can become a break-down path, allowing an electrical charge to flow between the earth and the cloud. A tall metal rod, with its base grounded to the earth can pass such an electrical charge harmlessly to earth, and will shield a small region around itself from lightning.

Three centuries ago none of these mechanisms were known. Sometimes a tall spire would be demolished by lightning, but others, perhaps with fortuitously grounded metallic sheathing, seemed never to be harmed. Explanations of all these separate phenomena were not tied together, but were given mostly in terms of many supernatural causes. It is now easier to learn and apply our modern explanation, than it was then to become familiar with all of the ancient folklore on the subject.

**1.2**   Note that the position of the size of human beings in Fig. 1-2 is about one-third the way between one meter and ten meters. That doesn't seem right, at first. Aren't most people closer to two meters tall? The explanation lies in the nature of the logarithmic scale. Two, on such a scale, is one-third of the way between one and ten. Three is about half-way.

The earth is eight thousand miles in diameter; that's

$$(8000)(8 \text{ miles}/5 \text{ Km}) \cong 13,000 \text{ kilometers} = 1.3 \times 10^7 \text{ meters}.$$

1.3   We would be scaling a meter down so that it is only the diameter of an atomic nucleus. That's quite a change in scale! Nevertheless, to represent a distance of $10^{26}$ meters, it would be necessary to use a map that is $10^{26}$ m $(10^{-15}$ m$/1$ m$) = 10^{11}$ m long. That is about the distance from the earth to the sun!

1.4   The triangle formed is a right triangle with the 90° angle at one observing station. The base of the triangle is 1000 miles, and the leg is 240,000 miles. The apex angle (near the moon) can be found from a simple trigonometric definition: $\tan \theta =$ opposite/adjacent $= 1000/240{,}000$. A trig. table gives the value of approximately $\frac{1}{4}°$ for this value of the tangent.

Alternatively, note that this very narrow triangle is approximately a thin slice of a circle with an arc of 1000 miles and a radius of 240,000 miles. The apex angle, $\theta$, is a small fraction of 360° and is given by:

$$\theta° = 360° \frac{\text{arc}}{\text{circumference}}$$

$$= 360 \frac{1000}{(2\pi)(240{,}000)} = 0.24°$$

For the same problem with the sun, $\theta°_{\text{apex}} \approx \frac{2}{3} \times 10^{-3}$ degrees, or about 2 seconds of arc!

1.5   By making two observations six months apart in time, our base line is equal to the diameter of the earth's orbit. The triangle formed by such measurements has a base line of 186,000,000 miles. The legs of the triangle (the distance to the nearest star) are $2 \times 10^{13}$ miles. From the map in Fig. 1-2, you can see that the equivalent distances in meters are: $3 \times 10^{11}$ m and $3 \times 10^{16}$ m. The apex angle is equal to about 2 seconds of arc.

Telescopes cannot be aimed with this precision. Instead, pictures of the same section of the sky are taken six months apart. The two negatives are placed over each other in careful alignment. Most of the stars are so far away (the "fixed stars") that their positions match exactly. An angular difference of position of about a second of arc is enough to place each of the images of close stars at slightly different positions on the two negatives. In order to measure their separation—and thus the distance to the star—a microscope must be used!

1.6   If you choose color as a classifying scheme, the boundaries would certainly be fuzzy. Furthermore, many objects, both animate and inanimate, change color over a period of time. Can you classify things growing in the ground or on bushes as fruits or vegetables without running into contradictions? What is a tomato? How about the eggplant?

1.7   If you say that a needle is a tool used in sewing, your definition is unsatisfactory. If you say that a needle is a thin, pointed, piece of steel you have left out its behavior. The definition must include both its appearance and its use.

Clearly, if you define a martini only in terms of its contents and do not describe its use and consequences, your definition is poor and possibly irresponsible.

1.8    Gravitation is a long range force. The strength of its influence depends on the masses of the objects involved and on the inverse square of the distance between the objects. When the fundamental particles are in actual "contact" (within distances of $10^{-15}$ meter), the nuclear forces are stronger than the gravitational by a factor of $10^{39}$! But the nuclear forces fall to zero strength at distances slightly greater than $10^{-15}$ meter. The gravitational force between a nuclear particle in your body and one at the center of the earth is infinitesimally small—only $10^{-78}$ newton or $2 \times 10^{-79}$ pounds! Nevertheless, there are about $10^{29}$ nuclear particles in your body and about $10^{52}$ of them in the earth. Each particle in your body is attracting each particle in the earth. Your weight is therefore:

$$10^{-78} \left( \frac{\text{newtons}}{\text{particle}} \right) (10^{29})(10^{52}) \approx 10^3 \text{ newtons} \approx 200 \text{ pounds.}$$

handling the phenomena

1.    We have all learned in school that the planets go around the sun. However, if you have ever identified Venus, Mars, Jupiter, or Saturn in the sky, it was probably not obvious that they and we circle the sun. Even if you observe that the planets move slightly night by night with respect to the stars, you would have a hard time relating such motions to elliptical orbits. Given the Copernican model, however, it is easier to work the other way and explain the motions that are actually observed.

You can make a scale model of the solar system that will display features most people have never seen. The distances between planets is so great compared with the diameters of the planets that it is difficult to make or observe a model that is accurately scaled for *both* diameters and orbital distances. For instance, if the earth is made only as large as a pea, the distance from earth to sun must be about 50 meters (about 150 feet). In Table 1.1 we list the diameters and orbital radii of the planets. To comprehend the size of the whole system, forget the planetary sizes and make only their orbital distances to scale. Choose a scale so that the whole solar

table 1.1

| Sun, Moon, and Planets | Diameter in Kilometers | Distance to Sun in Kilometers |
|---|---|---|
| Sun | 1,400,000 | — |
| Mercury | 5,000 | 58,000,000 |
| Venus | 12,000 | 110,000,000 |
| Earth | 13,000 | 150,000,000 |
| Mars | 6,800 | 230,000,000 |
| Jupiter | 140,000 | 780,000,000 |
| Saturn | 120,000 | 1,400,000,000 |
| Uranus | 48,000 | 2,900,000,000 |
| Neptune | 45,000 | 4,500,000,000 |
| Pluto | Up to 7,000? | 5,900,000,000 |
| Moon (ours) | 3,500 | (from Earth) 390,000 |

system can fit into a room, or at most, down a hall. For instance, if 1 cm is made equivalent to 5,000,000 km, Pluto will be almost 12 meters from the sun. The sun itself can be shown to scale on such a map, but it will be only 3 mm in diameter. The rest of the planets can be represented by marker tabs.

One convenient way of making the model is to use a roll of cash register or adding machine paper tape. Paste or tape a tab for the sun at one end. Mark out appropriate distances along the tape for the orbital radii of the planets. Fasten paper tabs at each of those marks and label them. With such a model you can roll up the whole solar system and put it in your pocket.

Notice how close the inner planets are to the sun, but how far away the others are. Is it not remarkable that such a small object as the sun can control the motion of Pluto at such a large distance?

2.   Using the information from Fig. 1-2 you should be able to construct scale models of two other realms—the atom and our galaxy. If you can, choose your scales so that the entire model is visible, but choose scales that will also show crucial individual features. In other words, show the nucleus of the atom in the first model, and the distances between individual stars in the galaxy model.

3.   The triangulation method of measuring great distances is best understood by actually using the system. Tape two protractors at the ends of a meter stick as shown in Fig. 1-10. Look across the sighting pin in the center of one protractor and line up the second pin with an object about 10 meters away. Leaving the meter stick stationary, sight on the same object using the pins and protractor at the other end. In the arrangement shown, each of the angles, $\alpha$ and $\beta$, will be about 3°. Their sum is the apex angle of the acute triangle that is formed with the object at the acute angle opposite the base formed by the meter stick.

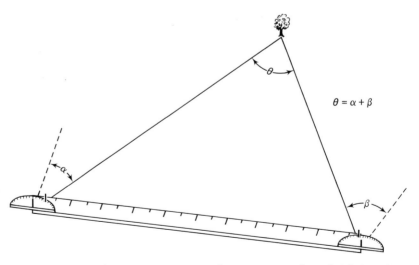

$$\theta = \alpha + \beta$$

figure 1-10   Two plastic protractors taped to a meter stick, and sighting pins that can be made from straight pins or paper clips, become a rangefinder.

Rather than using right angle trigonometry, it is more convenient with small angles to use the approximation that the baseline is an arc of a large circle. The distance to the object is the radius, r, of this circle. If $\theta$ is the apex angle, then:

$$\frac{\theta^\circ}{360^\circ} = \frac{arc}{circumference} = \frac{baseline}{2\pi r}.$$

Therefore the distance to the object, is:

$$\frac{\theta}{360^\circ} = \frac{baseline}{2\pi r}$$

$$2\pi r\theta = (baseline)(360^\circ)$$

$$r = \frac{(baseline)(360^\circ)}{2\pi\theta}$$

$$r \approx (baseline)\left(\frac{57^\circ}{\theta}\right)$$

Use your triangulation rangefinder to measure the distance to several objects. What is the longest distance you can measure without having the measurement uncertainties mask the results?

4.   According to folklore, Columbus challenged the scientific wisdom of his time by assuming that the earth was round. As a matter of fact, the Greek astronomers at Alexandria two thousand years ago not only knew that the earth was spherical but had measured its circumference. You can do the same thing, using the method of Eratosthenes, shown in Fig. 1-11. It was known that at noon on the summer solstice (about June 21—the longest day of the year), the sun shone straight down a deep well at a place on the Nile River now marked by the Aswan Dam. This point is on the Tropic of Cancer and, on June 21$^{st}$ of each year, the sun is directly overhead at noon. About 500 miles to the north, in Alexandria at the mouth of the Nile, Eratosthenes measured the angle made by the shadow of a vertical post at that same time. The angle of 7.2° is 7.2/360 of a full circle. Therefore the distance of 500 miles

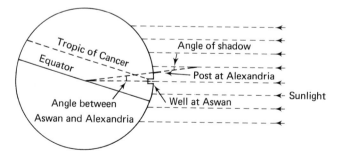

figure 1-11   The geometry used by Eratosthenes to measure the circumference of the earth.

table 1.2    latitude of various cities in the U.S.

| | | | |
|---|---|---|---|
| Anchorage, Alaska | 61.5° | Miami, Florida | 25.7° |
| Atlanta, Georgia | 33.7° | Milwaukee, Wisc. | 43.1° |
| Baltimore, Md. | 39.3° | New Orleans, La. | 30.0° |
| Boston, Mass. | 42.4° | New York, N.Y. | 40.9° |
| Chicago, Illinois | 42.0° | Portland, Maine | 43.6° |
| Cleveland, Ohio | 41.8° | Portland, Oregon | 45.6° |
| Dallas, Texas | 33.0° | St. Louis, Missouri | 38.7° |
| Denver, Colorado | 39.7° | San Francisco, Cal. | 37.5° |
| Detroit, Michigan | 42.5° | San Juan, Puerto Rico | 18.0° |
| Honolulu, Hawaii | 21.0° | Seattle, Washington | 47.5° |
| Houston, Texas | 29.6° | Tallahassee, Florida | 30.5° |
| Indianapolis, Ind. | 39.9° | Washington, D.C. | 39.0° |
| Los Angeles, Cal. | 34.0° | Tropic of Cancer | 23.5° |

must be 7.2/360 of the earth's circumference, C. If: $7.2/360 \, C = 500$ miles, then $C = 25,000$ miles.

To reproduce Eratosthenes' measurement, choose a day close to the equinoxes (Sept. 21st or March 21st) or solstices (Dec. 21st or June 21st). At noon on the equinox days the sun is directly overhead, on the equator. At noon on the winter solstice (December 21st) the sun is directly over the Tropic of Capricorn in the southern hemisphere. Table 1.2 gives the latitude for various cities in the United States. Every degree of latitude is about 70 miles along the earth's surface. Find the distance between your location and the latitude where the sun is directly overhead at noon of the day on which you make your measurement. Find the angle cast by a shadow of a vertical post. (Solar noon is not necessarily clock noon; you may be on daylight saving time. Furthermore you might not be on the central meridian (a line of longitude in your time zone that is evenly divisible by 15). Watch the shadow during an hour before and after the hour. Solar noon occurs at the time when the shadow is shortest.) Define the "arc" as the distance in miles between your location and the point where the sun is directly overhead at noon. Then:

$$\frac{\text{(your measured angle)}}{\text{arc}} = \frac{360°}{\text{(earth circumference)}}.$$

problems

1    Describe your own body measurements in terms of the metric system. If you cannot obtain metric scales or meter sticks, translate the measurements from the English system (which England no longer uses).

1 meter = 100 centimeters = 39.4 inches = 3.28 feet
1 kilogram weighs 9.8 newtons or 2.2 pounds.
height _____ cm
weight (mass) _____ kg
waist _____ cm

2    What is the metric equivalent of 60 miles per hour in terms of kilometers per hour?

3    You will be making some measurements of speed in terms of centimeters per second. How fast is 30 centimeters per second in terms of feet per second?

4    Water has a density of one gram per cubic centimeter. What is its density in pounds per cubic foot?

5    Estimate your volume in cubic meters. To do this, figure out the length, width, and height of the smallest box that you could fit in. Consider now whether or not your estimate is reasonable. What is the weight (mass) of a cubic meter of water, in kilograms? Since humans can just barely float in water, your density must be about the same as that of water. Assume that your estimated volume, in cubic meters, contains water. What would be the mass of the water in kilograms? How does that number compare with your weight?

6    The diameter of most atoms is about $2 \times 10^{-8}$ cm. *About* how many of them are lined up along one edge of a cubic inch of a solid? About how many atoms are in the cube?

7    The average distance from the earth to the moon is about 240,000 miles. With what *average* speed, in kilometers per hour, does a space ship have to travel in order to make the trip in two days?

8    A dime (about one centimeter in diameter) can obscure the moon if held at a distance of one meter from the eye. Similarly, the moon can just obscure the sun, as seen from the earth (thus we have eclipses). If the moon is $3.8 \times 10^5$ kilometers from the earth, what is its diameter? If the sun's diameter is $1.4 \times 10^6$ kilometers, how far is it from the earth?

9    Jet planes commonly cruise at altitudes of 30,000 feet. What is this altitude in kilometers?

10    Use a pan balance or postage meter scale to measure the mass of a penny, nickel, dime, and quarter in grams. (If you must use a scale calibrated in ounces, convert your answers to grams. 1 ounce = 28.4 grams)

*Chapter Two*

# simple rules for simple events

The universe is too large and too complex to be grasped as a whole. If we want to understand what it is and how it works, we must start with some small, simple part of it. Of course, if we had to work our way through the whole universe, small part by small part, the understanding would take a long time. Fortunately for our purposes, everything in the universe is made of only a few components, they interact with each other in only a few ways, and they are subject to only a few general laws. There is no guarantee that, if we know about these fundamental interactions, we will be able to figure out the way everything works together. But, that is a possibility and a hope. At least it seems reasonable that unless we can understand very simple events we cannot hope to understand ones that are complex.

As an example of this method of trying to simplify complex problems, consider how you might describe the path of a ball thrown through the air. It is standard to assume, as a first approximation, that the ball is not influenced by the air, that the gravitational force is constant, and that the influence of the earth's rotation is negligible. For many purposes those are good approximations. The resulting equation of motion is very simple. If the ball is a ping pong ball, however, such an approximation will probably be very poor. The ball will be strongly influenced by the air not only slowing it down, but also (if the ball is spinning) causing its path to curve. Projectiles traveling long distances would be influenced not only by the air, but by the changing gravitational field, and by the rotation of the earth. However, the complete calculations are so formidable that we would never understand them if we did not start out by learning the simple first approximation.

Choosing simple models for analysis is not always an obvious process. Such attempts in biology are sometimes called "reductionism." The word is used in a pejorative sense by some biologists who contend that knowing the details of molecular processes does not lead to an understanding of the whole biological system. Indeed, if the whole is more than the sum of its parts (or if the organizational scheme itself is the most important feature), it might not do any good to start studying the simple, individual parts. In this case, as in any other, the problem is to find the simple model.

## 2.1  behavior of a single object

What is the simplest phenomenon we can study? Surely it should have as few objects as possible interacting with each other. The smallest number is one. One object by itself cannot, by definition, interact with anything, but in this case we ought at least to be able to find out what the object itself does. However, isolating an object from everything else is not so simple. As a matter of fact, a completely isolated object is, in principle, unknowable. Unless it interacts with light or some other probe, we would never know that it exists.

*Question 2.1*    One of the authors owns a watch containing tiny Swiss watchmakers. They are the ones who keep the hands moving, but every time the case is opened they turn themselves into springs and gears. Are they real?

Even if we have reason to believe that our probes, such as light, do not have much influence on a particular large object, it is still difficult to isolate the object from everything else. We can shield it against air currents simply by putting up screens. Although, if we wish to remove all the influences due to the air, we would have to put the object in a vacuum system. To keep out heat we would have to enclose the object in insulation. External electrical influences can be stopped with a complete metal shield, although if we wanted to keep out magnetic fields we should make the shield out of iron. In some cases these shields must be complicated, consisting of several layers of different materials. No matter how complicated we make the shield, however, we cannot keep out gravity. There is no insulator, nor is there any known way to turn gravity off or deflect it. If we go sufficiently far away from the earth, the gravitational field will be small, although within the solar system we would still be influenced by the sun, which in turn is constrained by the gravitational field of the galaxy. There are ways, however, to compensate for gravity or to ignore it. One way to compensate for gravity is to provide a friction-free support for the object. It will still be subject to a constant vertical force due to gravity, but if the support is truly friction-free, at least the horizontal motion of the object will be free of any outside influence.

Friction-free supports are not easy to produce. As a first approximation we could use a cart with ball bearing wheels. The picture in Fig. 2-1a shows what happens when such a cart is given an initial velocity. It slows down and stops, which is hardly surprising! The technique for demonstrating this action will be used often in the next few chapters. It is called stroboscopic photography. In a darkened room, the camera shutter is opened during the entire time of the action and the objects are illuminated by the bright flashes of light from a strobe lamp. Each burst of light lasts about $\frac{1}{1000}$ of a second (or less, for some purposes). These flashes are repeated at a predetermined frequency. The light used to take the pictures in Fig. 2-1 was flashing at a rate of two times per second. If the cart moved 10 centimeters between flashes, its average velocity must have been 20 cm/sec.

A better method of reducing friction was used while the picture shown in Fig. 2-1b was being taken. The object, called a glider, was literally floating above the long track. Compressed air pumped into the hollow track at one end comes out through numerous tiny holes all along its surface. The glider is actually lifted about one tenth of a millimeter above the track's surface by the air pressure. The air does not drive the glider, but only supports it. When an object supported against gravity in this way is given an initial velocity, it continues to move at that same velocity. Of course, eventually it slows down because there is still friction caused by the air through which it moves.

*Question 2.2*    How good is the approximation to constant velocity of the glider in Fig. 2-1b? Measure and compare the distances traveled during the equal intervals of time.

Another way to compensate for the effect of gravity on an object is to ignore it, and let the object fall. This is a phenomenon that astronauts have

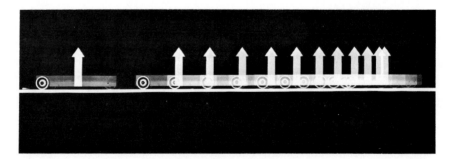

figure 2-1a    Strobe photo of a cart rolling toward the right and gradually slowing down. The cart is a block of wood mounted on four roller skate wheels. A vertical, white arrow has been mounted on the cart to indicate its position.

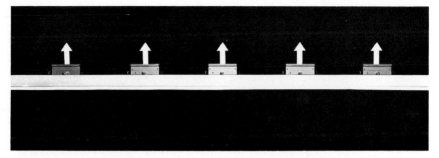

figure 2-1b   Strobe photo of a glider on an air track, moving toward the right
with almost constant velocity. A vertical, white arrow has been mounted on the
glider to indicate its position.

demonstrated in their television broadcasts to earth. When in orbit, or coasting
to or from the earth, the capsule is in a state of "free fall" so long as the
rocket engines are not on. The capsule and its contents are still subject to
the influence of gravity, but since they are all falling at the same speed they
are in a state of weightlessness. That is to say, a weight measuring device
(such as a bathroom scale) would register 0 if an astronaut were to stand on
it. Under such conditions an object released in the capsule will float freely
without any gravitational influence from the earth, *relative to the rest of the
capsule.*

   We can't produce quite the same effect in the laboratory on earth. However, we can record *horizontal* motion of objects while they are falling freely.
Gravitation affects only the vertical motion and so whatever happens in the
horizontal direction will be free of outside influence. The strobe picture in
Fig. 2-2 shows two balls that were launched horizontally from a table. They
fell, picking up speed in the downward direction, but were free to move in
the horizontal direction, each with its own constant velocity, without any
influence except the negligible air friction.

*Question 2.3*   Measure the horizontal displacements during each equal time interval.
What happened to the horizontal velocity?

   It appears that for an object moving horizontally, the more nearly friction-free it is, the more constant remains its speed. We can sum up the
observation by claiming that *an object free from other influences will maintain
constant velocity,* where velocity has the technical significance of representing
both a speed and a direction. Most of us were brought up believing that this
statement, called Newton's First Law, is true, and so we are more than willing
to believe it. Note, however, how extremely difficult it is to demonstrate the
law. We need clever gadgets to reduce friction and even these are only

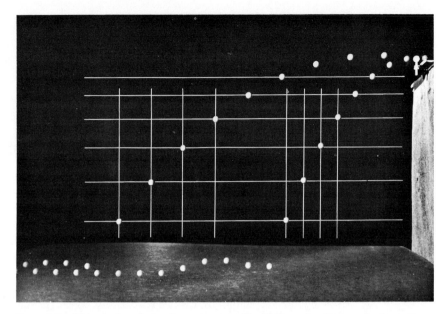

figure 2-2   This strobe photo shows two marbles falling simultaneously from the height of a table. One ball had been on a tee at the edge of the table and the other rolled down from the right and hit it head on. They both shot forward, moving to the left, each with a different horizontal velocity. Those velocities remained constant while their vertical velocities increased steadily and at the same rate. After they hit the floor, they bounced along toward the left. Vertical and horizontal lines were scribed on the picture afterward. The separation between vertical lines is proportional to the horizontal velocity. Note that this separation is different, but constant, for each ball. The separation between horizontal lines is proportional to the vertical velocity. Note that this separation increases steadily and is the same for both balls.

partially successful. The air track gliders do slow down and stop eventually. To measure the effect and thus be able to draw conclusions about what would happen without any friction at all, it is necessary to take strobe pictures or to record the motion with other sophisticated devices.

Far from being obvious, Newton's First Law would have been thought silly by most generations of human beings. They knew perfectly well that if they stopped pulling a cart, the cart stopped. To get something to move with a constant velocity, it was necessary to shove (or pull) it constantly. A famous problem in Greek philosophy, and one that was not solved until Galileo explained it in our modern terms, was why an arrow keeps moving after the bow string no longer propels it. Explanations were given in terms of the way the air, parted by the arrow tip, might flow around the arrow and push it from behind. With modern sophistication, we now turn the problem around and think that if an object slows down we should look for the source of friction.

*Question 2.4*  If a bicycle near us is standing still we may be willing to believe that there are no influences acting on it—at least, in any horizontal direction. Suppose someone, pedaling hard, rides past us at a constant velocity of 15 mph. Does that mean that the bicycle is free from influences?

There is one other problem associated with the behavior of an isolated object. Suppose that we first observe the object at rest, with zero velocity. We say that it is free from net outside influences. Then, without touching the object, let us watch while *we* move steadily past it at a constant velocity of 10 mph. (Almost everyone has had the experience of looking out of the window of a car or train which begins to move very gradually. It is difficult to tell whether the car is drifting forward or the outside world is moving backward.) Similarly, as our observing system moves to the right past the object, it will appear that the object is traveling at a constant velocity to the left with respect to us. Such a situation causes no trouble with our definition of an uninfluenced object. Both 0 mph and 10 mph are possible constant velocities, and so we can continue to maintain that the object is not subject to any net influences. But suppose we start accelerating past the object, either in a straight line direction or in the more complicated way shown in Fig. 2-3. Now the object is certainly not moving at a constant velocity, with respect to us. Are there outside influences acting on it? What started out to be a trivial definition turns out to be not so obvious, either to us or to the Greeks. We will maintain our definition by stressing its limitations: If, *with respect to an observer*, an object has constant velocity, then there are no net outside influences acting on the object, *so far as that particular observer is concerned.*

## 2.2   the interaction of two objects

What happens when the single object is subjected to outside influences? Let us once again tackle the simplest case and consider only what happens when *two* objects influence each other. For example, instead of asking how all the planets interact with each other and with the sun, it is far easier and more fruitful to find out how one planet and the sun would affect each other without the interference of anything else. The influence due to the other planets is fairly small and can be figured out as a correction to the first problem.

Such a "two-body" problem is not always a good approximation to a real event. A psychological study of how one person influences one other person may not explain what happens to a person in a mob where everybody influences everybody else. The motion of an electron in a television tube could, in principle, be calculated by considering the detailed interaction of that electron with every other separate electric charge in the tube, but it would

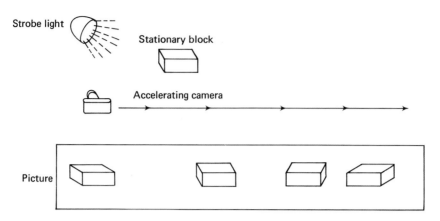

Strobe light

Stationary block

Accelerating camera

Picture

figure 2-3a   If a strobe photo of a stationary block were taken by an accelerating camera, the picture would be indistinguishable from that taken by a stationary camera of an accelerating block.

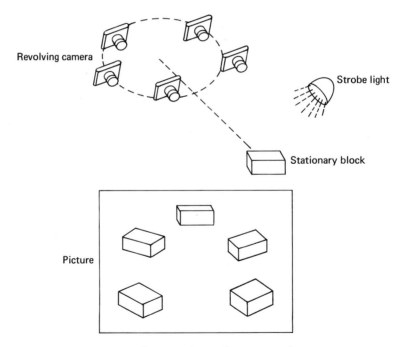

Revolving camera

Strobe light

Stationary block

Picture

figure 2-3b   In this case the strobe picture of a stationary block taken by an overhead camera revolving around a point slightly higher than the block will appear to be a picture of a revolving block.

not be a practical method. Still, there are many phenomena that can be explained by considering the interactions of just two objects. Such a study is worthwhile because it is simple, and therefore practical, and because the techniques that can be derived this way are powerful.

An interaction of two objects, uninfluenced by any others, is the most basic phenomenon that we can study. The "objects" can be large or small and can themselves be made up of many other objects. All that we require is that the object behave like a single system in our particular application. For instance, the sun is an enormous and very complicated object, and yet in studying planetary motion we can treat both the sun and a planet as simple individual point-like objects. The generalizing rules that we will find by studying the interactions of laboratory apparatus will apply to sun and planet, to the interaction of light with atoms, and to familiar objects in everyday use. (Fig. 2-4)

Two-body interactions can occur over various intervals of time; from the head-on collision between two cars, to the orbit of a planet around the sun. In the first case the strength of the interaction is essentially zero until the cars touch; in the second case the interaction strength depends on the separation distance in a simple, smooth and regular manner. Many interactions, such as the car collision, are very complicated if followed in detail. Even an apparently simple collision, such as that between billiard balls, involves forces and changes of shape at the moment of contact that would be very hard to analyze. Nevertheless, for some purposes it is easier to deal with interactions that begin and end in a short period of time. Regardless of how complicated the processes are during the interaction, some general simplifying statements can be deduced by examining the relationship of the system before the interaction to its condition afterwards.

*Question 2.5*    For the simplest case to study, we should examine two objects whose actions are uninfluenced by any others and which influence each other only for a brief time. Does a collision of one ball rolling against another

figure 2-4  In this electronic flash picture, taken at the moment of collision of a golf club with the ball, the compression of the ball is clearly shown. An instant later, the compressed ball will spring away from the club. The details of the interaction at the moment of collision are very complicated. However, very simple generalizations can be made to describe the final results of such a two-body interaction. (Flash picture courtesy Professor Harold Edgerton, Massachusetts Institute of Technology.)

satisfy these conditions? Study the stroboscopic picture of such a collision shown in Fig. 2-5, and cite evidence for your conclusion.

The collision of rolling balls is much too complicated for our simple purpose. The table plays an important role, not just in compensating for gravity, but in determining the horizontal motions of the balls. Friction constantly slows the balls and at the moment of impact they must dig into the surface slightly as they change their rolling speed and direction. Furthermore, the motion of each ball is not simple but is a combination of rolling and forward movement.

Ideal objects for the study of simple interactions would glide without friction and have no rolling motion. We can approximate such conditions by using the laboratory carts or floating gliders shown in Fig. 2-1. Their use permits another simplification—only motion in one dimension is allowed. Two simple types of interactions between pairs of objects can be provided by springs or by magnets lined up to repel each other as shown in Fig. 2-6. The range of interaction is longer with magnets than with the springs, but it is still short compared to the travel lengths of the carts or gliders. The advantage of using these interactions, rather than simply allowing the objects to bump into each other, is that no change takes place in the objects themselves because of the interactions. There are no dents and the temperature does not rise.

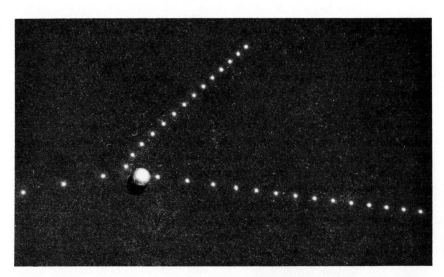

figure 2-5    Strobe photo of collision between large ball bearings on a floor. The target ball was stationary and made the large white spot at the collision point since, in that position, it was illuminated by several flashes. Meanwhile the other ball shot in from the left. The target ball rolled forward after the collision, and the incoming ball was deflected to the left of its original direction.

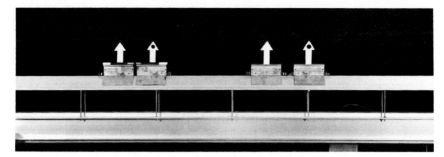

figure 2-6   Two pairs of air track gliders. The pair on the left interact because cylindrical magnets are mounted on them, with opposing poles facing one another. The pair on the right interact because they are equipped with spring bumpers. (The magnets must be separated from the track by a vertical distance of several centimeters so that there is no "damping" effect caused by an interaction of the moving magnets with the aluminum track.)

Such interactions are called elastic, and the everyday meaning of the word describes the conditions hoped for. In Chapter 6 we shall define the conditions more definitely and be concerned with interactions that are not elastic.

Before examining photographs of the glider collisions, recall the line of reasoning that has led to such an inquiry. In order to explain stars, dinosaurs, rainbows, and lovers we have turned to an examination of the simplest interactions we can produce. Gliders on an air track may seem remote from the real world, but unless we can explain simple interactions we should not expect to understand complex ones. Note, however, how hard it is to produce simplicity. When Thoreau, referring to life style, said, "simplify, simplify," he was asking for something very difficult. Similarly with physical phenomena, it is easy to roll one ball against another producing a complex interaction, but difficult to devise a two-body collision free from outside influences. Once we have achieved the simple interactions our aim is to be able to describe them as simply as possible. We want *generalizations* about such interactions. If we have to provide a separate description for each different case, we have made no progress at all.

In looking for generalizations it might seem reasonable to specify how things change during an interaction. After all, that's the crux of an interaction; things change. There's another possibility, however. Perhaps during interactions some things never change. Keep this possibility in mind as you examine the photographs of two-body collisions.

Understanding the rest of this chapter depends crucially on your observations and measurements of the pictures in Fig. 2-7. At the end of the chapter there is a suggestion about how you can duplicate the pictured events with an air track. You might want to take your own strobe photos for analysis, but it is difficult to get precision results without elaborate equipment. At any rate, study the sequence of pictures until you understand the special conditions demonstrated in each one as explained in this section. Velocity measurements can be made directly from the pictures with the use of a ruler. The strobe light flashed every half second.

Notice, first of all, the interactions between two *identical* gliders, as shown in the first four pictures. In the first one (2-7a) the gliders start from rest and are shoved apart by a spring that had been compressed between them. In the second picture (2-7b) one glider is initially at rest and the other one collides with it through the interaction of the spring. The next two pictures duplicate the first two situations respectively, except that in Figs. 2-7c and 2-7d the interaction is provided by the mutual repulsion of magnets carried by the gliders. In all four cases the gliders can be considered identical simply because they are the same size and are made out of the same material.

The next three photographs are of interactions between a glider with unit size and a glider *twice* as big. In 2-7e the two have been shoved apart by an expanding spring. In 2-7f the double glider has collided with the unit glider, which had been motionless. Fig. 2-7g illustrates just the opposite situation; the unit glider has collided with the double glider, which had been motionless.

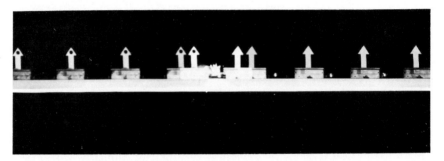

figure 2-7a   Two identical gliders, originally at rest in the middle of the track, are shoved apart by a spring between them.

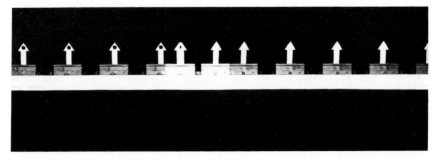

figure 2-7b   The two gliders are identical. The one with the plain arrow (on the right) was initially at rest in the middle of the track. The other came from the left and struck the first one with a spring. The target glider moved off to the right while the bombarding glider stopped.

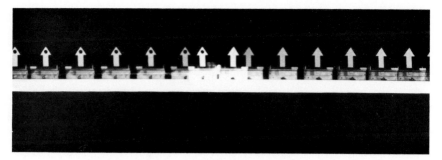

figure 2-7c  Two identical gliders, initially at rest in the middle of the track, are shoved apart by a pair of opposing magnets, one of which is fastened to each glider.

figure 2-7d  The two gliders are identical. The one with the plain arrow (on the right) was initially at rest in the middle of the track. The other came from the left and the magnet mounted on it repelled a similar magnet on the target glider. The target glider then moved off to the right while the bombarding glider stopped.

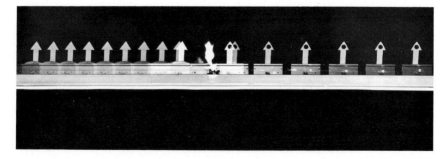

figure 2-7e  The two gliders were initially at rest in the middle of the track and were then shoved apart by a spring between them. The glider on the left was twice the size of the one on the right.

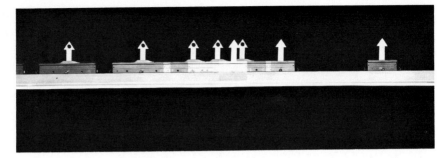

figure 2-7f   The unit glider (with plain arrow) had been at rest in the center of the track. The double glider came in from the left, shoving the unit glider forward to the right, and continued on to the right itself with reduced velocity.

figure 2-7g   The double glider (with the plain arrow) had been at rest in the center of the track. The unit glider came in from the left, shoved the double glider forward to the right, and bounced back with reduced velocity. Note the direction indicator on the arrow of the unit glider. While traveling to the right the indicator is swept back as though it were a gun across a man's shoulder. In the collision the indicator flips to the right, so that once again it is "over the shoulder" of the arrow as it moves to the left.

figure 2-7h   The two gliders were initially at rest in the middle of the track and were then shoved apart by a spring between them. The glider on the left was three times the size of the one on the right.

figure 2-7i   The unit glider (with the plain arrow) had been at rest in the center of the track. The triple glider came in from the left, shoved the unit glider forward to the right, and continued on to the right itself, with reduced velocity.

figure 2-7j   The triple glider (with the plain arrow) had been at rest in the center of the track. The unit glider came in from the left, shoved the triple glider forward to the right, and bounced back with reduced velocity. Once again, the direction indicator on the arrow of the unit glider is pointed back "over the shoulder" of the arrow as it advances.

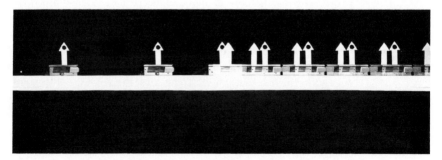

figure 2-7k   Two identical gliders were used for this collision, but this time there was a sticky material between them instead of a spring or opposing magnetic field. The collision was nonelastic. One glider (with a plain white arrow) was initially at rest in the middle of the track. The other one (with a black dot on its arrow) came in from the left. When they collided, they stuck together and moved on to the right.

The sequence for the next three pictures is like the previous sequence except that a *triple* glider now interacts with a unit glider. In 2-7h a spring has shoved them apart from rest; in 2-7i the triple glider struck a motionless unit glider; in 2-7j the unit glider struck a motionless triple glider.

In all of the pictures described above, great care was taken to make sure that the interaction itself did not interfere with the gliders in any permanent way. The collisions were "elastic", as defined earlier, since they did not leave the gliders dented or warmer. Fig. 2-7k, shows a completely non-elastic (or inelastic) collision. The two gliders stuck together after the collision, necessarily compressing and disturbing the sticky material that bound them together.

# 2.3   simple rules for special cases—momentum

It would be very difficult to describe each of the pictured events in terms of the variable compression of the spring or the changing velocities of the gliders during the moments of interaction. If we were to attempt such a description, we would have to provide a different one for each different type of interaction. For instance, the gliders equipped with magnets influence each other over a longer distance than those with the springs. Our goal is to describe each event in terms of conditions before and after the event takes place. We can either find those properties that *change,* or observe any characteristics that *do not change. The surprising discovery is made that the changes taking place in simple events of this type are severely restricted by laws that require certain properties to be unchanged.* If we can spell out these laws, we can greatly simplify the description and understanding of even complicated events.

What remains constant in the events with identical gliders? Of course, the gliders themselves do not change shape, size, or color. All three of these properties might have changed if there had been gunpowder between the gliders instead of springs. In these simple events, however, only the gliders' velocities seem to be affected. Remember that with only one object, the defining characteristic of "no net external influence" was constant velocity. *The collision events of Figs. 2-7b and d also yield constant velocity—of the two-glider system as a whole.* To be sure, the velocity of the bombarding glider is transferred to the target glider, which takes off with the original velocity, while the first one abruptly stops. With this slight variation of meaning, however, we could claim that *the total velocity of the system has been conserved.*

However, this circumstance does not completely explain what happens. If only conservation of original velocity were required, the two gliders might both have continued on together in the same direction, each having half of the original velocity. Yet in elastic collisions, this never happens. Furthermore, it appears at first glance that velocity is not conserved at all in the explosion events of Figs. 2-7a and c. In these events the original velocity was zero, yet obviously each glider has velocity afterwards.

*Question 2.6*    How could velocity be defined so that it could be conserved in these special cases?

Even with an expanded definition, velocity is not conserved in the events shown in Figs. 2-7e and h. It is not obvious that anything about these motions is the same both before and after the explosions. In the first case the original velocity was zero. The smaller glider shot off to the right with twice the speed of the larger one which went to the left. In the second case the smaller glider took off with three times the speed of the larger one. These simple speed ratios provide a clue to a method for inventing a conserved quantity. Remember that when identical gliders were shoved apart, each acquired a speed equal to that of the other but opposite in direction. In our first case of nonidentical glider interactions, the larger glider is exactly twice as big as the small one. The gliders are all made out of the same material, have the same cross section, but the larger ones are two or three times as long as the unit glider. Let us define a property, which we shall call *inertial mass* $(M)$, that is a characteristic of an object related to the way its velocity changes. For our particular purpose, we can define another quantity, which we shall call *momentum*, as the product of velocity and inertial mass. Momentum $\equiv Mv$. We can arbitrarily propose that momentum is conserved in the explosion events. This claim defines the measurement of the inertial mass of any object. For instance, in Fig. 2-7a the original momentum of the two-glider system was zero. The momentum after the explosion was: (unit inertial mass) $(+ 52 \text{ cm/sec}) +$ (unknown inertial mass) $(-52 \text{ cm/sec})$. If we insist that momentum be a conserved quantity, the momentum after the explosion must also be zero. The unknown inertial mass is therefore equal to the unit inertial mass. This event is shown schematically in Fig. 2-8.

*Question 2.7*    Using the same argument, find the defined inertial masses of the larger gliders in Figs. 2-7e and h.

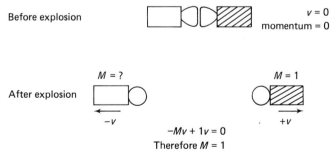

figure 2-8   Schematic illustration of a method of measuring the inertial mass of an unknown object.

To summarize, the argument is this: by assigning direction to velocity we can claim that velocity is conserved in the interaction of *identical* objects. When non-identical objects interact, velocity is not conserved. In order to maintain a conservation law, we invent a quantity, momentum, which is a product of the velocity of an object and some characteristic of its size and the material from which it is made. This characteristic, which we call inertial mass, is defined in such a way that momentum is conserved when two objects, with masses $M_1$ and $M_2$, mutually repel each other.

Momentum before explosion = Momentum after explosion

$$0 = (M_1 v_1) + (M_2 v_2)$$

Choose a unit inertial mass by letting $M_1 = 1$. Then, any other inertial mass, $M_2$, can be measured by letting it be repelled by $M_1$. The value of $M_2$ will be: $M_2 = -(v_1/v_2)$.

There is no guarantee that there will always be conservation of momentum with such a definition, and with inertial masses measured in this way. If the law and definitions do not apply to many other cases, there is no point to the whole procedure. Let us try out the rules on the rest of the collisions shown in Fig. 2-7. In Fig. 2-7f, the glider with an inertial mass of 2 units collides with a glider having a unit inertial mass. In Fig. 2-7g the same two gliders collide again, but this time the glider with $M = 2$ was originally at rest. In Figs. 2-7i and j there are similar collisions between gliders with $M = 1$ and $M = 3$.

*Question 2.8*  Check the data from these pictures to see whether or not momentum is conserved.

It might appear that we have defined momentum so that it *has* to be conserved. That is almost, but not quite, the situation so long as we deal with the defining action of two objects repelling each other. Suppose we use our procedure and find that object #2 has 2 units of inertial mass compared with our unit inertial mass, object #3 has 3 units, etc. Logic does not at all require that if objects #2 and #3 repel each other, their relative velocities will be 3:2. It turns out experimentally that this is the case, and it makes life very convenient—but it was not logically required by the original definitions. We are so accustomed to the fact that inertial masses are added together as if they were ordinary numbers that it seems both obvious and natural that they do so. Other quantities, such as volume and velocity, do not necessarily combine in the same obvious way.

*Question 2.9*    Give an example of how the volume of two objects combined is different from the sum of the individual volumes. Give a similar example for combined velocities.

Momentum conservation becomes more a "law", and less a *definition*, as it is successfully applied to more and more phenomena. As you could see from answering Question 2.8, momentum was conserved in *all* the collisions that were photographed. In the next chapter we will extend the range of application by considering two-dimensional events. We will also see that momentum is conserved not only with laboratory toys but with other objects ranging in size from sub-atomic particles to planets. While so far we have only been analyzing two-body interactions, we will find that the momentum of a system is conserved no matter how many objects are in the system. There is an important qualifying point here, however. The system must be isolated from all outside influences—remember how we went to great trouble to isolate the two interacting gliders. The objects included in the system might be interacting with each other in many ways, but so long as the system as a whole is not being influenced by conditions outside itself, the momentum of the whole system is conserved. If the momentum of a system consisting of two gliders is initially zero, then the momentum of the system (though not of the individual gliders) remains zero after they explode from each other. If, however, some outside influence acts on the gliders, the momentum of the system will depart from this value.

The property of an object called its inertial mass, $M$, was defined so that momentum would be conserved in two-body explosions. This definition provided an operational method for measuring the inertial mass of an unknown object. One feature of the definition is that the measurement can be done anywhere—on earth, in space, on the moon. (No gravitational field is necessary; in fact, the presence of a gravitational field must be compensated for.) This circumstance might not seem to be much of an advantage, but note how different the procedure is from measuring the weight of an object. We normally think of mass as being related to weight, and mass measurements usually consist of weighing procedures. Why do we propose the recoil method of measuring mass, and why call it "inertial mass $(M)$", instead of just "mass, $m$"? The answer is that, from what we have seen so far, there is not the slightest reason to assume that inertial mass has anything to do with that property of matter called gravitational mass, which is proportional to an object's weight. The first quantity has something to do with the "reluctance" of an object to change its velocity; the second quantity has something to do with influencing other objects from a distance. The first quantity is measured by timing the motion of objects in a changing situation; the second quantity is measured with entirely different apparatus in a static situation.

In Chapter 4 we will define gravitational mass, *and will then demonstrate the astonishing fact that inertial mass is proportional to gravitational mass.* This is an experimental finding of profound importance. We dare not miss the full significance of the relationship by casually assuming that the word "mass" applies to both inertial and gravitational properties.

## 2.4   kinetic energy

We started out looking for conserved quantities in two-body interactions— quantities that are the same after an interaction as they are before. The product of inertial mass and velocity is such a quantity. Perhaps other combinations of mass and velocity are also conserved. Momentum would lose its peculiar importance if $(Mv)^2$, $(M^2v)$, $(Mv)^3$, and other combinations are the same before and after an interaction. Note that there is no reason to expect that they will be, because we are concerned with the *sum* of these quantities. For example, suppose that the initial momentum of the bombarding object $(Mv)_o$ has the value, 6. After the collision the bombarding object and the target object could both be traveling in the same direction, each with a momentum of 3.

$$(Mv)_o = (Mv)_1 + (Mv)_2$$

$$(6)_o = (3)_1 + (3)_2$$

*Question 2.10*    The numbers above could occur if an object with an inertial mass of 3 units and a velocity of 2 collides with an object with a unit inertial mass. In this case the bombarding object continues on with a velocity of 1 while the target object attains a velocity of 3. Does $(Mv)_o^2 = (Mv)_1^2 + (Mv)_2^2$?

Tests for conservation of various other combinations of $M$ and $v$ can be tried with data from Fig. 2-7. The first four interactions pictured there are between identical gliders and do not provide good tests because the identical inertial mass cancels out of every term. Let us analyze only the four cases in which there was some original velocity of one of the gliders: Figs. 2-7f, g, i, and j. For each case determine the various masses and velocities and fill out a chart like the one that follows. Do not forget that velocity has direction and that our convention is to label it positive if it is toward the right and negative if it is toward the left. Velocity squared is always positive, however, since the square of a negative number is positive.

| $M_1$ | $M_2$ | $v_o$ | $v_1$ | $v_2$ | $(M_1v_o)$ | $M_1v_1$ | $M_2v_2$ | $(M_1v_1 + M_2v_2)$ |
|---|---|---|---|---|---|---|---|---|
| | | | | | $(M_1^2v_o)$ | $M_1^2v_1$ | $M_2^2v_2$ | $(M_1^2v_1 + M_2^2v_2)$ |
| | | | | | $(M_1v_o^2)$ | $M_1v_1^2$ | $M_2v_2^2$ | $(M_1v_1^2 + M_2v_2^2)$ |
| | | | | | $(M_1^2v_o^2)$ | $M_1^2v_1^2$ | $M_2^2v_2^2$ | $(M_1^2v_1^2 + M_2^2v_2^2)$ |
| | | | | | $(M_1^3v_o^3)$ | $M_1^3v_1^3$ | $M_2^3v_2^3$ | $(M_1^3v_1^3 + M_2^3v_2^3)$ |

*Question 2.11*   For the four cases shown in Fig. 2-7f–j compare combinations of $M$ and $v$ before and after collision. Is there any combination that is conserved?

The particular combination of $M$ and $v$ that is conserved is different in many ways from momentum, $(Mv)$. In the first place, since it is necessarily positive it has no direction, positive or negative, and in the next chapter we shall see that it has no direction at all. Momentum, on the other hand, always has a specific direction. Furthermore, momentum is conserved in many circumstances where $(Mv^2)$ is not. Look at Fig. 2-7k. In this simple case (a nonelastic collision) the identical gliders stuck together when they collided and continued on with half of the original velocity. Momentum was conserved but $(Mv^2)$ was not, as you should demonstrate for yourself. In all four of the explosion cases, $(Mv^2)$ was not conserved although, by definition, momentum was conserved. In these cases the original velocity and therefore the original $(Mv^2)$ was zero. Since both gliders had positive $(Mv^2)$ afterwards, there could be no cancellation and therefore no conservation of $(Mv^2)$. It almost appears as if the occasional conservation of $(Mv^2)$ would have very little importance. Rather than give it up, however, we can propose that it is a quantity that can be turned into other forms. The game would then be to discover these other forms and see if we can invent enough of them to satisfy a general conservation law. That procedure has been the history of understanding of this quantity during the past century. The name assigned to $(Mv^2)$ is kinetic (or motion) energy. For consistency with other uses, kinetic energy is defined as: $E_{kin} = \frac{1}{2}Mv^2$. The factor of $\frac{1}{2}$ makes no difference in analyzing two-body *elastic* collisions, but enters into any situation where kinetic energy is transformed into other forms of energy. In Chapters 4, 6, and 7 we shall investigate some other properties of objects that we can define as other forms of energy.

## 2.5    applications of the conservation laws

Even the limited conservation of kinetic energy can be very useful in analyzing the interactions to which it applies. These include the collisions shown in Figs. 2-7b, d, f, g, i, and j (all the elastic collision cases). Momentum conservation alone is not sufficient to explain the particular velocities observed. For instance, if only momentum conservation were required, the identical glider collisions might have resulted in both gliders continuing on together with half the original velocity. That is just what happened in the inelastic collision shown in Fig. 2-7k and momentum was conserved in that case. In the elastic cases, where there was no sticking or denting, $(\frac{1}{2} Mv^2)$ was conserved but this happened only if the bombarding glider stopped completely while sending the identical target glider on with the bombarding glider's original velocity. To see that this behavior is *required*, we will write down the two conservation laws:

Conservation of momentum, $(Mv)$:

$$M_1 v_o = M_1 v_1 + M_2 v_2 \qquad (2\text{-}1)$$

Conservation of $E_{kin}$, $(\frac{1}{2} Mv^2)$:

$$\tfrac{1}{2} M_1 v_o^2 = \tfrac{1}{2} M_1 v_1^2 + \tfrac{1}{2} M_2 v_2^2 \qquad (2\text{-}2)$$

Since $M_1 = M_2$, we can simplify the equations by dividing through by the value of the common mass, and also by dividing the energy equation by $\frac{1}{2}$:

$$v_o = v_1 + v_2 \qquad (2\text{-}3)$$

$$v_o^2 = v_1^2 + v_2^2 \qquad (2\text{-}4)$$

Square both sides of equation (2-3):

$$v_o^2 = v_1^2 + v_2^2 + 2v_1 v_2 \qquad (2\text{-}5)$$

Compare the squared equation with equation (2-4), the energy equation. They agree only if $(2v_1 v_2) = 0$. If that is the case, either $v_1$ or $v_2$ must equal zero. What we observed in Figs. 2-7b and d was that the bombarding glider stopped, that is $v_1 = 0$. Substituting this value for $v_1$ into equation (1), we find that the velocity of the target glider, $v_2$, is equal to the original bombarding velocity, $v_o$. That is exactly what happened.

*Question 2.12*    What about the other possibility? The algebra would agree with a situation where after the collision a bombarding glider would have velocity $v_1 = v_o$, and the target glider would have velocity $v_2 = 0$. How can this be?

The two conservation equations completely determine the final velocities of an elastic collision. There are only two unknowns—$v_1$ and $v_2$. Two independent equations that are simultaneously true can always be solved for two unknowns. As an example, let us predict the final velocities of the gliders pictured in Fig. 2-7g. In that case a unit glider with $M_1 = 1$ ran into a glider with $M_2 = 2$.

Conservation of momentum, $(Mv)$:

$$1v_o = 1v_1 + 2v_2 \tag{2-6}$$

Conservation of $E_{kin}$, $(\frac{1}{2} Mv^2)$:

$$(\tfrac{1}{2})(1)v_o^2 = (\tfrac{1}{2})(1)v_1^2 + (\tfrac{1}{2})(2)v_2^2 \tag{2-7}$$

Simplifying the equations:

$$v_o = v_1 + 2v_2 \tag{2-8}$$

$$v_o^2 = v_1^2 + 2v_2^2 \tag{2-9}$$

Squaring equation (2-8):

$$v_o^2 = v_1^2 + 4v_2^2 + 4v_1v_2 \tag{2-10}$$

Subtract equation (2-9) from equation (2-10):

$$0 = 2v_2^2 + 4v_1v_2$$

Divide through by $2v_2$:

$$0 = v_2 + 2v_1$$

Solving for the final velocity, $v_1$:

$$v_1 = -\tfrac{1}{2}v_2$$

The final velocity of the bombarding glider is half that of the target glider and in the opposite direction. Substitute that value for $v_1$ back into equation (2-8). It appears that $v_2 = \tfrac{2}{3}v_o$. The final velocity of the target glider was two-thirds that of the original bombarding velocity and in the same direction. Those are, indeed, the results obtained from Fig. 2-7g.

*Question 2.13* Apply the laws of conservation of momentum and kinetic energy to the situation shown in Fig. 2-7f where the glider with inertial mass 2 strikes the glider with inertial mass 1. Do the final velocities required by the equations agree with the observed results?

## summary

We have turned from marveling at the size and complexity of the universe to the consideration of the simplest events we can find. The hope is that the laws governing simple phenomena will have more general validity. First we asked what the motion of an object would be if all external influences were removed. Our answer defines the condition of the observer as much as the behavior of the object. If there is no net external influence on an object, as determined by an observer, then the velocity of an object remains constant with respect to that observer. Such a statement can equally well define what we mean by saying that an object is not subject to any net influence with respect to us; if the velocity of the object with respect to us is constant, we conclude there is no net influence.

The next simplest event is the two-body interaction. Its analysis is fundamental to the understanding of most other reactions, because often we reduce these reactions to two-body cases by making averages or approximations. In this chapter we studied only two-body interactions along one dimension. Air-track gliders were used to approximate a situation where the two-body system would be free from external influences. We looked for quantities that retained the same numerical value after the interaction as they had before it.

With identical gliders, velocity was conserved so long as plus and minus directions were assigned to velocity. To maintain a conserved quantity in nonidentical glider interactions, we defined a new quantity called momentum. Momentum is the product of the inertial mass of an object and its velocity. The inertial mass is defined in such a way that momentum is conserved in a two-body explosion event. These arbitrary definitions gain significance when it appears that momentum is conserved in all the two-body interactions. Another quantity involving velocity and the inertial mass is also conserved in certain two-body interactions which are characterized as being "elastic." This quantity, $\frac{1}{2}Mv^2$, is called kinetic energy. Momentum has direction and can be positive or negative in the one dimensional case, but kinetic energy has no direction and is always positive. These two laws are concerned with things that do not change during an interaction. Surprisingly enough, they so restrict the changes that *can* take place that it is possible to use the laws to predict the final velocities of any two-body, one dimensional, elastic interaction.

*Answers to*
*Questions*

2.1 So long as the Swiss watchmakers can never be detected, it makes no difference whether or not they are there—but they had better keep the watch running.

2.2 To answer this question, and also Question 2.3, you must actually make measurements on the photographs. If possible, use a transparent plastic ruler calibrated in centimeters.

2.3   See above.

2.4   We should only claim that the bicycle is free from any *net* influence. The friction and drag effects must be just canceled out by the effect of pedaling, thus allowing the observed constant velocity.

2.5   Notice in the strobe picture that the path of the bombarding ball is not quite straight immediately after the collision. Since it must roll in the direction that it is moving, and since this direction changes at the moment of collision, the ball has to skid slightly and so dig into the surface in order to start rolling in the new direction. Some slowing down of the balls is also evident.

2.6   We should assign a directional marker to velocity. Mathematically, in these one-dimensional motions we can call a velocity to the right positive, and a velocity to the left negative. Since the magnitude of the velocity of each glider is equal to that of the other member of the pair, the sum of the velocities remains zero. $(+v - v = 0)$

2.7   Fig. 2-7e:

Momentum before = momentum after $(M_1 = 1)$

$$0 = (M_2)(-5.0) + (1)(+10)$$

Therefore,

$$M_2 = 2$$

Fig. 2-7h:

$$0 = (M_3)(-3.5) + (1)(+10.5)$$

Therefore,

$$M_3 = 3$$

In this measurement and all others with Fig. 2-7, it is not necessary to know the scale between the printed photograph and the original track, nor is it necessary to know the strobe light frequency. Since the scale and frequency remain constant, the velocity of any object shown is *proportional* to the *distance* it moved between flashes. Simply measure that distance with a ruler on the printed photograph, and use that number (in millimeters) as the velocity. Note that in these equations, the proportionality constant given by the scale factor cancels out.

2.8   Momentum is indeed conserved, as your measurements should demonstrate.

2.9   A liter of water added to a liter of ethyl or methyl alcohol yields less than two liters of alcohol-water mixture. A plane flying at 100 mph in a cross wind of 30 mph does not travel at 130 mph.

**2.10**  No, $(6)^2 \neq (3)^2 + (3)^2$

Consider also the possibility of the equality:

$$(M^2v)_o \overset{?}{=} [(M^2v)_1 + (M^2v)_2]$$
$$(3)^2(2) \neq (3)^2(1) + (1)^2(3)$$

Consider also any of the other combinations. Are there any that are conserved?

**2.11**  One, and only one, of the combinations is conserved.

**2.12**  Nothing in the algebra requires an actual collision. Although on the track the gliders cannot pass through each other, the same equations do describe a situation where one object simply passes by the target. In that case the target would keep its zero velocity and the bombarding object would maintain its original velocity.

**2.13**  Conservation of momentum:

$$(2)v_o = (2)v_1 + (1)v_2 \tag{2-11}$$

Conservation of $E_{kin}$:

$$\tfrac{1}{2}(2)v_o^2 = \tfrac{1}{2}(2)v_1^2 + \tfrac{1}{2}(1)v_2^2 \tag{2-12}$$

Rewriting these equations:

$$2v_o = 2v_1 + v_2 \tag{2-13}$$
$$2v_o^2 = 2v_1^2 + v_2^2 \tag{2-14}$$

Solve (2-13) for $v_2$:

$$v_2 = 2v_o - 2v_1 \tag{2-15}$$

Square equation (2-15):

$$v_2^2 = 4v_o^2 + 4v_1^2 - 8v_o v_1 \tag{2-16}$$

Substitute $v_2^2$ into equation (2-14):

$$2v_o^2 = 2v_1^2 + 4v_o^2 + 4v_1^2 - 8v_o v_1 \tag{2-17}$$

Divide by 2 and consolidate terms:

$$3v_1^2 - 4v_o v_1 + v_o^2 = 0 \tag{2-18}$$

Apply the general solution for quadratic equations and solve for $v_1$:

$$v_1 = \frac{4v_o \pm \sqrt{16v_o^2 - 12v_o^2}}{6}$$
$$= \frac{4v_o \pm 2v_o}{6}$$
$$= v_o \text{ or } \tfrac{1}{3}v_o$$

Find $v_2$ by substituting these values in equation (2-13):

$$v_2 = 2v_o - 2v_1; \text{ or } 2v_o - 2(\tfrac{1}{3}v_o) = \tfrac{4}{3}v_o$$
$$= 2v_o - 2(v_o) = 0$$

### handling the phenomena

The basic data needed in this chapter are provided in the precision photographs of Fig. 2-7. By following the instructions given on page 45, you can obtain measurements directly from the pictures. However, it would be much easier to understand the events if you could actually see them take place. If you have an air track and gliders the effects are easy to reproduce, at least qualitatively. Try duplicating each of the events shown in Fig. 2-7.

If you have time and the equipment, you could duplicate the photographs yourself. You need a strobe light, or preferably two strobe lights in order to illuminate the whole track evenly. The background should consist of dark cloth. It is a tricky business to produce the right lighting conditions and then to stage the collision events with the correct timing. For the explosion events, fasten the gliders together with thread. Separate them by cutting the thread with a razor blade. (If you try the traditional method of burning the thread, the air from the air track will probably blow out the match.) Make sure that the track is level along its whole length and that there is sufficient air pressure to support the gliders if you should load them with bar magnets. Finally, make sure that your camera and its location do not produce distortions in the length measurements from one end of the track to the other. (The pictures shown were taken by college students—it can be done!)

### problems

1  One of the simplest of two-body elastic collisions occurs when the target object, $B$, is originally at rest and is then struck head-on by an *identical* object, $A$. As the strobe pictures show, the target object takes off with the original speed, $+V_o$, of the bombarding object, while the bombarding object itself stops dead. See what your description of such an event looks like from different reference frames.

(a)  Suppose that you, the observer, are moving along beside the bombarding object, $A$. Your velocity, $V_o$, continues constant. What velocity do you observe for object $A$ before the event? What velocity do you observe for $B$ before the event? What are the velocities of $A$ and $B$ after the event? (In answering, be sure that you use $+$ for velocities to the right, and $-$ for velocities to the left.)

(b)  Suppose that you are traveling in the same direction that $A$ has originally but with velocity, $+\tfrac{1}{2}V_o$.

What velocities do you observe for $A$ and $B$ before and after the collision?

2  An object with mass four, initially at rest, is struck by an object with mass one. The bombarding object was traveling from the left with velocity, $V_o$. What are the final velocities of each? Remember that both kinetic energy and momentum must be conserved.

3    Two objects, originally at rest on a friction-free track, shove apart from each other. Object *A* obtains a velocity of 35 cm per second, and object *B* goes at 7 cm per second. Object *A* has a mass of *two*. What is the mass of *B*?

4    Object *A*, with a mass of two, runs into object *B*, which is initially at rest and has a mass of three. The original velocity of *A* was +10 cm per second. The final velocity of *A* was −2 cm per second, and the final velocity of *B* was +8 cm per second. See if these values satisfy the principles of conservation of momentum and kinetic energy.

5    In another collision of the objects described in Problem 4, the original velocity of *A* was +10 cm per second. The final velocity of *A* was −0.5 cm per second, and the final velocity of *B* was 7 cm per second. Was momentum conserved? Was kinetic energy conserved? Is the effect a possible one?

6    In another collision of the objects described in Problem 4, the original velocity of *A* was 10 cm per second. The final velocity of *A* was −4 cm per second and the final velocity of *B* was +7.5 cm per second. Was momentum conserved? Was kinetic energy conserved? Is the effect a possible one?

7    Object *B* is initially at rest on a frictionless track. Object *A*, with mass of two units, runs into it with an initial velocity of 10 cm per second and keeps going forward with a velocity 3.7 cm per second. Object *A* is knocked forward with a velocity of 12.6 cm per second. What was the mass of object *B*? Was the collision elastic?

8    Find out what fraction of the original kinetic energy is lost in an identical object collision when the two objects stick together after the collision. Assume that object *A* with velocity, $V_o$, runs into object *B* that is initially at rest. The mass of object *A* equals the mass of object *B*. After the collision the velocity of *A* equals the velocity of *B*, since they stick together.

9    Object *A*, with a velocity of 10 cm per second, strikes object *B* which was initially at rest. The collision is elastic with no kinetic energy being lost. The mass of object *A* is *much* larger than the mass of object *B*. Consequently, object *A* continues on its way after the collision with its original velocity scarcely diminished. What is the final velocity of object *B*? (First, write down the equations for conservation of momentum and kinetic energy, calling the final velocity of *A*, $V_A$, and the final velocity of *B*, $V_B$. Do not yet assume that $V_A \approx V_o = 10$ cm/sec. Next, square the equation for momentum conservation, and subtract it from the equation for energy conservation. You will obtain a relationship between $V_A$ and $V_B$. It is at this point that you can make the approximations that $m_B/m_A \approx 0$ and $V_A \approx V_o$.)

10   If a ball bounces off a solid wall, it appears that momentum is not conserved. The ball clearly changes momentum since it maintains about the same speed, but reverses direction. Change of momentum equals the final momentum minus the initial momentum:

$$-mv_o - (+mv_o) = -2mv_o.$$

Since the wall does not appear to move, it seems that the original momentum has changed by $-2mv_o$.

Analyze a collision between an object $A$ with a very small mass compared to that of object $B$. The initial velocity of $A$ is $+V_o$, and object $B$ is initially at rest. Write down the conservation equations. Square the momentum equation and subtract it from the energy equation. Let $m_A/m_B$ go to zero, and find the values for $V_A$ and $V_B$. Note that it is only in the approximation when $m_A/m_B \longrightarrow 0$ that $V_B = 0$.

# two-dimensional interactions

The regularities in collisions that we observed in the previous chapter have limited usefulness. To be sure, two-body interactions are common and basic to our understanding of the physical world, but usually the action is not constrained to take place only along a line. In this chapter we will examine the usefulness of our definitions of momentum and kinetic energy in two-dimensional interactions. Instead of events taking place only along a line, we will allow the colliding objects to move freely in a plane.

Once again we need experimental techniques that will allow us to eliminate, ignore, or compensate for gravity. There are air tables, similar in operation to the air track that we used in Chapter 2. Compressed air leaks up through thousands of tiny holes, supporting floating disks. These are very useful for demonstrating qualitative effects, but it is hard to obtain quantitative collision data with them. Instead of compensating for gravity in this way, we can ignore it by allowing the collisions to take place in three-dimensional space. That is what is happening in the marble pictures shown in Fig. 2-2, reproduced again as Fig. 3-1. The marbles are falling, but this vertical motion is completely independent of the horizontal motion. When starting from a position of rest, all objects fall with the *same* increasing velocity. Except for air resistance, the downward acceleration does not depend on the size or mass of the object. If the bombarding marble is traveling in the horizontal direction to begin with, both it and the target marble will strike the floor at the same time. To compare horizontal velocities, simply compare the horizontal distances traveled during the equal times of fall. The geometry of this arrangement

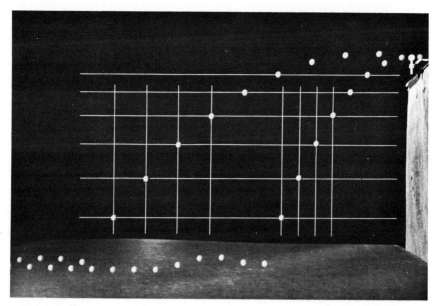

figure 3-1

is illustrated in Fig. 3-2. Instead of interfering with the interaction between the marbles, gravity provides a convenient timing method.

*Question 3.1*    Drop an object from table to floor and estimate the time it takes to fall that distance. Suppose a marble shoots off the table in a horizontal direction and hits the floor one meter from the edge of the table. Using your estimated time of fall, what was the horizontal speed of the marble?

At the end of this chapter there are directions and suggestions for obtaining two-body collision data. The apparatus is very simple and the precision of results can be high. As we did in one-dimensional events, we investigate first of all what happens when the two objects are *identical*. Obtain data for at least three collisions. One should be a head-on collision, or close to it, and the other two should be glancing collisions where the target marble goes off at an angle between 30° and 60° from the forward direction. In each case find the speeds in arbitrary units before and after collision. Since it is not necessary to use the actual speeds in the calculations, you can find the speeds in convenient arbitrary units just by measuring the horizontal distances traveled. For each of the marbles, the horizontal speed, $v_H$, is equal to the ratio of the horizontal distance traveled, $\Delta x$, to the time it took, which is the time of falling, $\Delta t$: $v_H = \Delta x/\Delta t$. The time of fall, $\Delta t$, is the same for both marbles.

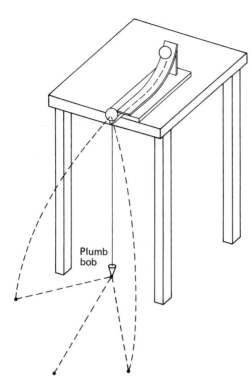

figure 3-2   Marble collision at the edge of a table. The landing spots are recorded on paper placed on the floor and covered with carbon paper. The extent of the line directly forward from the track, and the spot at the farthest end of it are determined by allowing the bombarding marble to shoot off without a collision.

Plumb bob

*If* our one-dimensional conservation laws hold, we might expect to find:

$$Mv_o = Mv_1 + Mv_2$$

and

$$\tfrac{1}{2}Mv_o^2 = \tfrac{1}{2}Mv_1^2 + \tfrac{1}{2}Mv_2^2$$

The speed of the bombarding marble before collision, $v_o$, is found by letting the marble shoot off the launcher without any collision. For convenience, call the speed of the bombarding marble after a collision $v_1$, and the speed of the target marble after the collision $v_2$.

*Question 3.2*   Test your data from your three trials and see whether or not these momentum and kinetic energy expressions hold. Notice that the equations can be simplified by cancelling out the common inertial mass, $M$, and by cancelling the common factor $\tfrac{1}{2}$ in the energy equation.

We might expect that the momentum law would not hold, because even in one dimension, momentum has direction—plus or minus. Somehow we must provide for that direction property in the more complicated two-dimensional case. The velocities, or the momenta, can be specified in terms of their magnitudes in the forward direction (the original bombarding direction) and the direction perpendicular to that. Notice the geometry of this description in Fig. 3-3.

*Question 3.3*    Test your data and see whether the original forward momentum is equal to the final forward momentum, and the original perpendicular momentum is equal to the final perpendicular momentum.

Once again it may seem as if we are manufacturing conservation laws—if our original definition does not work, alter it until it does! If we had to do that for each new case, it would not be worth the trouble. As we shall see, however, our expanded definition of momentum applies to many other phenomena.

There is one very special feature of identical object collisions in two dimensions. When the interaction is truly elastic there is always an "opening" angle of 90° between the two tracks after the collision. The geometry is illustrated in Fig. 3-4. This situation occurs regardless of whether the bombarding marble ends up going nearly forward or nearly perpendicular to its original direction. The presence or absence of this 90° angle makes a very sensitive test of whether a collision was elastic, or subjected to unknown errors. Check your own observations to see if your data agree with this requirement. At the end of the chapter on page 74 there are two derivations of the 90° condition, one algebraic and one geometric.

The collisions that we have talked about so far must be studied with special laboratory apparatus. Considerable care has to be taken to minimize friction and to ensure that the collisions are elastic. Normally, when two objects bounce

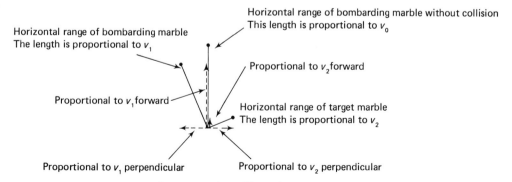

figure 3-3    Geometry of a collision between two marbles. Each of the actual ranges can be broken up into a forward component and a perpendicular component.

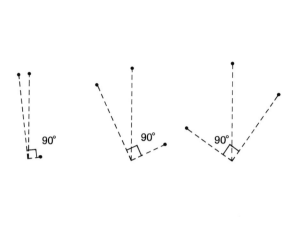

figure 3-4 Three different glancing collisions of identical marbles. Regardless of the angle of deflection from the forward direction suffered by the bombarding marble, the angle between the lines connecting the ranges of target and bombarding marble to the point directly under the origin is always 90°. This condition provides a sensitive test of whether or not the collision was elastic and the marbles identical.

off each other, there will be very slight denting and some of the kinetic energy will disappear. We will study that effect in Chapter 6.

Another source of friction-free, elastic collisions, undisturbed by gravity, is the interaction of sub-atomic particles. Their flight over short distances is so rapid that the effect of gravity is unmeasurable; in many instances their collisions are completely elastic, and if the particles have high energy, they can travel long distances through gases without losing very much energy to processes equivalent to "friction." Pictures of the trails left by such particles in a cloud chamber are shown in Fig. 3-5. A cloud chamber is a closed container of gas completely saturated with water vapor (100% humidity). A sudden expansion of one side of the chamber reduces the temperature abruptly. Less water can be in the form of vapor at the lower temperature. However, unless "seed" centers such as dust particles are present, rain drops will not form and grow. If, at this moment, an electrically charged sub-atomic particle shoots through the gas, it will leave a trail of damaged gas atoms in its wake. These atoms have been "ionized"—one or more electrons have been torn off each one, leaving an ion with a positive electrical charge. Both the positive ions and the negatively charged electrons can serve as the "seeds" for droplet formation. The ion trail turns into a droplet trail, which appears white against the black background.

In the cloud chamber picture shown in Fig. 3-5a, an electron coming from the left collided with an electron of one of the gas atoms. The target electron tore off, leaving its own trail, while the bombarding electron bounced in another direction. There is no easy way to measure the momentum of each electron in this picture (although it is possible with a different experimental arrangement), but note that the opening angle between the two final paths is 90°. As we have noted, this is characteristic of an identical body elastic collision. In the cloud chamber photograph shown in Fig. 3-5b the bombarding particle is a proton, the positively charged nucleus of a hydrogen atom. It is traveling in hydrogen gas and runs into one of the stationary nuclei, providing a proton-proton collision. Once again, the 90° opening angle indicates that the two objects have equal mass. Another proton-proton collision is shown in Fig. 3-5c, but the angle between the final paths is not 90°. This last picture was taken with very high energy protons traversing a bubble chamber. In a bubble chamber there is a liquid (in this case, liquid hydrogen) with a temperature just below its boiling point. A sudden expansion reduces the pressure. This lowers the boiling point for the liquid until it is below the temperature of the liquid hydrogen in the chamber. Bubble formation, due to boiling, starts on the damaged atoms left along the path of electrically charged particles,

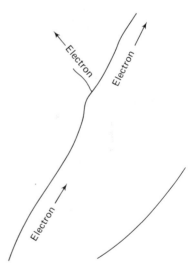

figure 3-5a   This is a rare picture of an electron-electron collision. A gamma ray source was placed near a low pressure cloud chamber. The gamma rays themselves leave no tracks, but they knock many electrons out of the walls and gas. In this case, one of these electrons traveled part way through the chamber and then suffered a glancing collision with an electron of one of the gas molecules. Since this was a collision of particles with equal mass, the angle between their subsequent paths was 90°. The electron paths were crooked because the electrons had low momentum and were easily scattered by the molecules through which they passed.

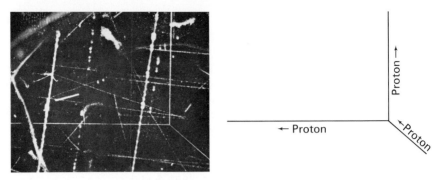

figure 3·5b  A proton traveling through a cloud chamber filled with hydrogen struck the nucleus of one of the hydrogen atoms. Since a hydrogen nucleus consists of just a single proton, the collision involved two particles with equal masses. The two protons went off at right angles to each other.

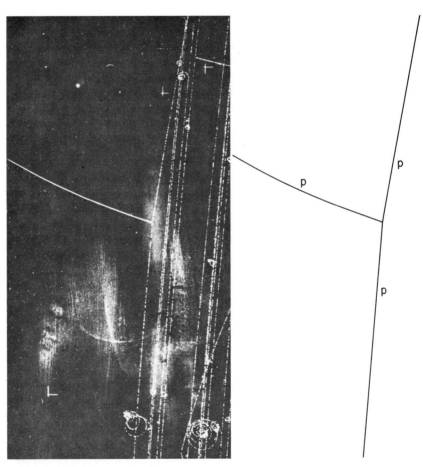

figure 3·5c  Proton-proton scattering at very high energy. This bubble chamber picture records the collision of a three billion electron volt proton with the nucleus of a hydrogen atom at rest. The relativistic mass increase made the bombarding proton almost four times as massive as the proton at rest. The resulting collision angle is less than 90°. (*Courtesy of Brookhaven National Laboratory, Upton, New York*)

which now form trails of bubbles. In this case, the bombarding proton was traveling close to the speed of light. The mass of an object at such a speed is greater than when that same object is standing still, or traveling slowly. Experimental proof for this relativistic increase of mass is provided by the bubble chamber picture. The proton-proton path angle is not 90°. Instead, it is the appropriate angle for a more massive particle striking a lighter one.

We must now test our laws of conservation of momentum and kinetic energy in the more general case of non-identical objects colliding in two dimensions. Use your marble apparatus to obtain data on several such collisions. For a head-on collision it will be necessary to use a lighter marble as the *target*. Both marbles will go forward in this case. If the bombarding marble is lighter it can bounce backward and run into the launch apparatus. Somehow you must find the ratio of masses of the two marbles. This can be done by measuring the relative speeds after a head-on collision between the heavy marble and the light one. There is an analysis of this situation at the end of this chapter on page 73. As shown there, the ratio of speeds after a head-on elastic collision between an object with mass $M_1$ striking an object with mass $M_2$ is:

$$\frac{v_1}{v_2} = \frac{1}{2}\left(1 - \frac{M_2}{M_1}\right) = \frac{M_1 - M_2}{2M_1} \tag{3-1}$$

A diagram showing sample data from a two body collision is shown in Fig. 3-6a. It is immediately apparent that the opening angle between the two final paths was not 90°. Since the collision was between hard marbles it must have been elastic. Therefore, the marbles must not have identical masses. Indeed, in this particular case, the bombarding marble had twice the mass of the target marble. Note that the original forward speed does not equal the arithmetic sum of the two final forward speeds, nor are the perpendicular speeds after the collision equal. It is not *velocity* that is conserved, but *momentum*. We must change the figures to forward and perpendicular momenta, and then compare these. The transformation is illustrated in Fig. 3-6b and tabulated below.

$M_1 = 2 \quad M_2 = 1 \quad v_o = 4 \quad M_1 v_o = 8$

$v_1 = 2.3 \quad M_1 v_1 = 4.6 \quad v_2 = 4.6 \quad M_2 v_2 = 4.6$

$v_1(\text{forward}) = 2 \quad M_1 v_1(\text{forward}) = 4$

$v_1(\text{perpendicular}) = 1.15 \quad M_1 v_1(\text{perpendicular}) = 2.3$

$v_2(\text{forward}) = 4 \quad M_2 v_2(\text{forward}) = 4$

$v_2(\text{perpendicular}) = 2.3 \quad M_2 v_2(\text{perpendicular}) = 2.3$

*Question 3.4*    Test these data for conservation of momentum and kinetic energy. Then analyze the data from your own experiments in the same way.

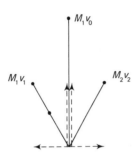

figure 3-6a    Sample data drawn to scale for a collision between a bombarding marble with mass two and a target marble with mass one. Note that the included angle, $v_1 O \, v_2$, is
not $90°$, and that the perpendicular velocities do not cancel.
The target marble velocity, $v_2$, is actually greater than the
original velocity, $v_o$, of the bombarding marble. The velocity
of the bombarding marble after the collision is $v_1$.

figure 3-6b    This is a momentum diagram, a conversion of the data in Figure 3-6a. Each of the velocity
displacements has been multiplied by its appropriate
mass, $M$. The displacement corresponding to $v_2$ has
been multiplied by 1, since the target marble had
unit mass. Both of the other displacements have been
multiplied by 2, since the bombarding marble had
twice the unit mass: $M_1 = 2$.

In this case, and in your own experiments (if all went well), momentum
and kinetic energy were conserved. Remember, however, both the restrictions
that apply to these cases and also the astonishing success of the original
definitions. As for the warning note, these interactions are strictly between
two bodies, uninfluenced by any third object. The gravitational interaction
of these objects with the earth, as a third body, was canceled out; and, friction
was almost eliminated. Most important, the collisions were elastic with no
internal change of the objects (such as heating or denting) to absorb some
of the kinetic energy. To observe how hard it is to obtain these special simple
conditions, look again at the ball-bearing collision in Fig. 2-5. Momentum
and kinetic energy are not conserved in that event.

The success of the original definition is clear if you remember how it
appeared that we were manufacturing the conservation laws. Inertial mass
was defined by insisting that when multiplied by velocity, the two formed
a conserved quantity called momentum. The inertial mass of each object could
thus be determined by recoil experiments between an unknown object and
a mass defined to be the unit mass. Momentum was further defined to have
direction; and, in such a way that its components parallel and perpendicular
to the original bombarding direction were conserved. The remarkable feature
is not that these definitions can be made; but, that having been made for simple
cases they turn out to be universally applicable. All combinations of objects
with various masses, colliding elastically at different angles, obey the conservation laws of momentum and kinetic energy. Floating disks, marbles,
planets, and sub-atomic particles are all subject to the same laws.

# 3.1   some applications of the conservation laws

If you fire a large caliber gun you will feel an immediate result of the law of conservation of momentum. See Fig. 3-7. The bullet, with small inertial mass, $m$, will shoot out at a high velocity, $V$, carrying away the momentum, $mV$. Since the original momentum of the whole system—bullet, gun, and you—was zero, it must remain zero. That can happen only if the inertial mass of the remaining system—you and the gun—recoils backward with a velocity, $v$. The actual inertial mass of the system is complicated because you are coupled to the earth by friction, and the gun is coupled to your shoulder, which might travel back faster than the rest of you. The kinetic energy for all this activity is produced by the gunpowder, a process that we will examine more closely later.

Rocket propulsion (as well as all other kinds of propulsion) depends on the conservation of momentum. The rocket blast does not shove back against the air in order to push the rocket forward. In fact, rockets work better in a vacuum, where there is no air to get in the way. The process is almost the same as if a machine gun were mounted on the back of a sled—or a spaceship. The bullets would always leave the gun with a specific velocity, $V$, relative to the gun. With respect to the ship, each bullet carries a momentum equal to $mV$, going in the backward direction, which must be exactly compensated by an equal change of momentum of the remaining system going in the forward direction. The more bullets that are shot out in this way, and the greater their velocity, $V$, the more momentum will be added to the main system. Rockets use molecules instead of bullets, and the velocities can be no greater than the velocity of the molecules in exploding gases—about 1000 meters/second. See Fig. 3-8.

*Question 3.5*   Don't rockets travel faster than this? How can they if the exhaust gas cannot go back any faster?

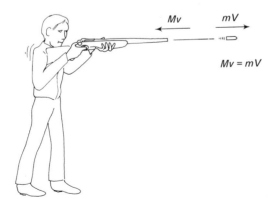

$Mv$    $mV$

$Mv = mV$

figure 3-7   The recoil of a gun must be taken up by the assemblage of shoulder, man, and earth. The momentum of the bullet is the product of its small mass, $m$, and large velocity, $V$.

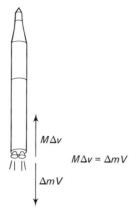

$M\Delta v$

$M\Delta v = \Delta m V$

$\Delta m V$

figure 3-8   The rocket is driven by momentum recoil. For a given small mass, $\Delta m$, of gas shooting out from the tail end with a high velocity $V$, the rocket (with mass $M$) gains the velocity $\Delta v$.

To gain sufficient speed to go into orbit or escape the earth, rockets using chemical fuels must shoot a major fraction of their mass out the back end. If the outgoing velocity of the gas could be increased, less would have to be shot out in order to produce the same thrust. One way to do this might be by accelerating the molecules with strong electric fields, in the same manner that some types of atom smashers accelerate particles. Such a method might be used for deep space probes, but it cannot be used for the initial launch in the atmosphere because the particle acceleration must be done in vacuum.

Momentum conservation dominates the disintegration of an exploding shell or skyrocket. Since the exploding influences are all internal, the original momentum of the shell must be maintained as it breaks up. (Gravity continues to pull all the pieces downward, even as it was pulling on the original shell. The trajectory of the "center-of-mass," an average position of the pieces, does not change—except for air friction.)

*Question 3.6*    Suppose a shell explodes in mid-trajectory into two pieces of equal mass, and let us assume that one of them gets kicked backward so that all of its initial horizontal velocity disappears, and it falls straight down. See Fig. 3-9. If momentum must be conserved, what must be the velocity of the second piece? If the original mass of the shell equaled two units, and the original velocity was one unit, was kinetic energy conserved?

If the momentum of any object does change, there must be some outside influence acting on it. Suppose you are driving a car at high speed on an icy road and come to a curve. Originally, your momentum is large and aligned in the direction of travel. It would be desirable to change the momentum, either by slowing down, or by changing the direction of the car so that it matches that of the road. Only external influences will produce these effects,

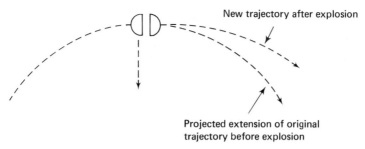

New trajectory after explosion

Projected extension of original
trajectory before explosion

figure 3-9   Shell bursting into two pieces in such a way that one
half is kicked backward, losing its velocity completely. It falls
straight down.

however. Your brakes must cause the tires to interact with the road, either
to slow down or to turn. The nature of the external influence is not always
obvious. The influence shoving your car sideways on the turn may be evident
in the skid marks on the road. A ball whirling in a circle is constrained by
a string not to fly off in a straight line with constant momentum. The planets
whirling about the sun leave no skid marks and are influenced by no visible
string. The influence is there, however. Gravitational pull between planet and
sun keeps changing the momentum. See Fig. 3-10. So pervasive is the momentum conservation law that whenever the momentum of an object changes with
respect to us, we ascribe the effect to some influence external to the object.

*Question 3.7*    A ball bounces off a wall, leaving with almost the same speed that it had
when it arrived. Has its momentum changed? Did the momentum of the
wall change?

Evidence about the behavior of subatomic particles is obtained from cloud
chamber and bubble chamber pictures. The pictures confirm that the laws
of conservation of momentum and energy apply to the microworld, and in
turn, the laws help us to interpret the pictures. In Fig. 3-11 a sequence of
particle decays occurs. A negative pi meson, $\pi^-$, a particle emitted from an
atomic nucleus that has been bombarded by high energy subatomic particles

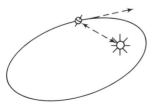

figure 3-10   Since the momentum of a planet
changes, there must be an external force acting on
it.

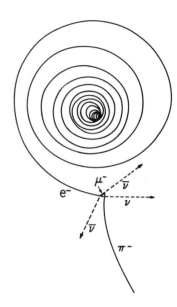

$$\pi^- \longrightarrow \mu^- + \overline{\nu}$$
$$\phantom{\pi^- \longrightarrow} \hookrightarrow e^- + \nu + \overline{\nu}$$

figure 3-11   In this bubble chamber picture, a negatively charged $\pi$ meson decays to a muon, which in turn decays to an electron. (*Courtesy of Brookhaven National Laboratory, Upton, New York*)

enters the chamber from the bottom. As the meson tears electrons away from the hydrogen atoms in its path, it loses energy and slows down, leaving a slightly thicker trail as it comes to rest. From other experiments we know that it takes about $10^{-8}$ seconds (one-one hundred millionth of a second) to decay after it comes to rest. It turns into a lighter particle called a negative muon, $\mu^-$, which shoots off with only a small kinetic energy. The muon trail is short and thick; its energy is rapidly lost in the process of ionizing the atoms along its path. About $10^{-6}$ second (one millionth of a second) after it stops it decays into an electron, e⁻. The electron has much less mass and so it does not take much kinetic energy to impart a high speed to it. Bubble formation along its trail is sparse, since it is going so fast that it does not stay near any atom long enough to have much of a chance to ionize it. All three of these particles—pi meson, muon, and electron—have a negative electric charge. A strong magnetic field was present in the bubble chamber. As is the case with all charged particles moving through a magnetic field, the paths of the meson, muon, and electron were curved. Only the path of the pi meson and the electron are long enough to show this effect. The curvature of the electron path is very dramatic because the electron has so little mass. The magnetic field changes its momentum easily, forcing the electron into an approximately circular path. Because the electron is losing energy and momentum all the time in passing through the bubble chamber liquid, the magnetic field continuously forces it into a tighter and tighter circle, producing a spiral path.

Even without knowing anything about particle physics, we can still see some surprising things in this picture. Look at the junction between the pi meson and the muon. The pi meson came to rest and then a short time later the muon shot out to one side. How can this be? If a hand grenade rolled to a stop, could it suddenly explode in such a way that all of it went dashing off in one direction? Momentum would not be conserved! There must be some other particle emitted in the direction opposite to that of the muon. Since electrically charged particles leave tracks, this recoil particle must have zero electric charge. (Such an arrangement would also be necessary to conserve electric charge. Both the negative pi meson and the resulting negative muon have an electric charge of the same sign. Therefore, the third particle must be neutral.)

This recoil particle is called an antineutrino, $\bar{\nu}$. It has no electric charge and can never remain at rest. Nevertheless, it still exists and plays an important role in subatomic particle interactions. The analysis of many pictures of this decay sequence reveals that the muon always gets the same energy from the decay. In the same bubble chamber its track is always the same length, although it can be in any direction. Such a constant energy is produced only in two-body decays; that is, the muon and neutrino always share the "explosion," or decay, energy in the same way.

*Question 3.8* Now look at the point where the muon decays and gives birth to an electron. What must be true about this event?

Another type of particle interaction is shown in Fig. 3-12. Negative pi mesons were being shot through a hydrogen bubble chamber. One of them struck the nucleus of one of the hydrogen atoms—a proton—and apparently just stopped. But, we know better! The original pi meson momentum must be somewhere. Further along in the same direction two pairs of tracks start

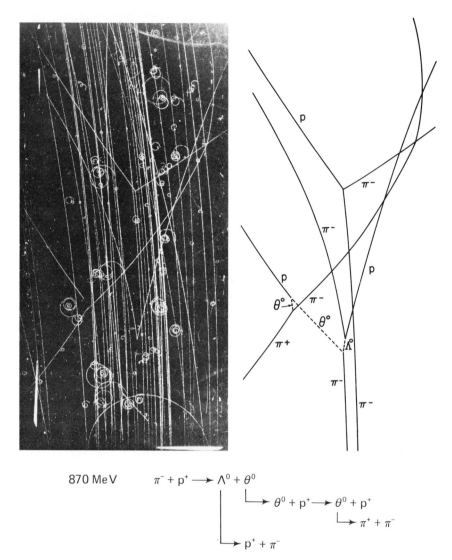

870 MeV        $\pi^- + p^+ \longrightarrow \Lambda^0 + \theta^0$

$$\hookrightarrow \theta^0 + p^+ \longrightarrow \theta^0 + p^+$$

$$\hookrightarrow \pi^+ + \pi^-$$

$$\hookrightarrow p^+ + \pi^-$$

figure 3-12   In this bubble chamber picture, a negatively charged $\pi$ meson struck the nucleus of a hydrogen atom. (The chamber was filled with liquid hydrogen.) Two neutral particles were produced in the collision. Since momentum is conserved, they travel in the same general direction as the original meson. After a short distance, one of the neutral particles ($\Lambda^0$) decayed to a positively charged proton and a negatively charged meson. The other neutral particle ($\theta^0$) struck a hydrogen nucleus and knocked it forward. The $\theta^0$ bounced back, having lost most of its kinetic energy, and then decayed into two mesons. (*Courtesy of Brookhaven National Laboratory, Upton, New York*)

out, seemingly from nowhere. All of this can be explained by assuming that the negatively charged pi meson joined the positively charged proton and turned into one or more neutral particles. Apparently there were two such particles, each of which decayed in flight (not at rest) into two other electrically charged particles. Each pair must have one negative and one positive particle in order to conserve electric charge. (Note here another of the important conservation laws; in all interactions, total electric charge must be conserved.) One pair consists of a negative pi meson and a positive pi meson, and the other pair is a positive proton and a negative pi meson.

*Question 3.9*   How can it be determined that the neutral particles decayed in flight?

So far we have defined momentum as the product of inertial mass and velocity. Some problems connected with this definition have already arisen. (Remember that in the high energy proton-proton collision, the incoming proton acted as if it had greater mass.) In the original example of the negative pi meson decay, we attempted to save the momentum conservation law by saying that a neutrino recoiled in the opposite direction from the muon. But we also said that the neutrino has zero mass. How can a particle with zero mass have momentum?

Figure 3-13 illustrates another example of this paradox. The cloud chamber was being irradiated by very high energy x-rays, which leave no trail because they have no electric charge. One of the x-rays turned all its energy into the mass of two electrons, positive and negative, and into their kinetic energy. After the "pair creation", the electron and its anti-particle, called a positron $e^+$, continue in the forward direction. But they are in a magnetic field, so their paths are curved. The negatively charged electron spirals in one sense (clockwise, from our viewpoint), and the positively charged positron spirals in the opposite sense (counterclockwise). In Chapter 7 we will consider some of the problems involved in mass-energy transformations. For now, consider only the momentum conservation situation. The x-ray is a chunk of electromagnetic radiation, similar to, but more energetic than light radiation. Clearly it had momentum, some of which it gave to the electron pair.

Our formula for momentum, $Mv$, is just an approximation which is good only for low velocities. For most purposes we do not have to worry about the complete formula unless the velocity is very close to the speed of light. Some subatomic particles can travel only at the speed of light. Their momenta are equal to $E/c$, where $E$ is their energy and $c$ is the speed of light. That expression successfully predicts the amount of momentum carried by x-rays and neutrinos.

For many purposes momentum is a more fundamental quantity than mass. It is often easier to measure the momentum of a particle directly than it is to measure its mass. For instance, the radius of curvature of the path of an

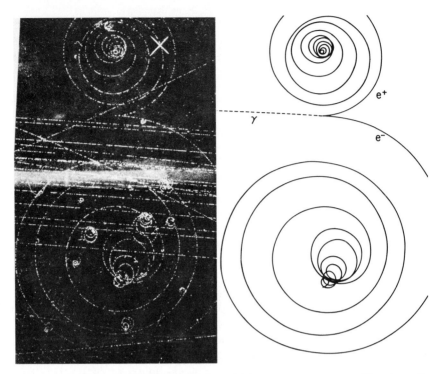

figure 3-13   Electron pair production by a high energy gamma ray. (*Courtesy of Brookhaven National Laboratory, Upton, New York*)

electrically charged particle in a magnetic field is proportional to its momentum. Double the momentum in the same magnetic field, and the particle will move in a circular path with twice the radius of its previous path.

*Question 3.10*    Compare the momenta of the electron and positron in Fig. 3-13 (shortly after they were produced). In Fig. 3-11, compare the momentum of the negative pi meson when it first entered the chamber to the momentum of the electron when it was first produced.

## 3.2   units and magnitudes

We have been describing mass and velocity in terms of arbitrary "units." This was satisfactory for most of our purposes because only relative comparisons of mass or velocity were needed. For future work it will be convenient to use standard units for all these quantities. Because the metric system is easier to use than the so called British units, and because the metric system is used in most scientific work, we will use it in this book.

The unit of length is the meter. Like most of the other standard units it is a convenient size for humans and is defined in terms of accurately measured references. It is no accident that it is about the same size as the yard; that also is a convenient "human size" unit. This particular feature of the units is emphasized in Fig. 3-14. The French committee that set up the metric system in 1791 chose to make the standard length equal to $1/10,000,000^{th}$ of a quadrant of the earth—$1/40,000,000^{th}$ of the circumference. This strange and completely arbitrary choice produced a standard of length that is useful for measuring household objects and, in the form of meter sticks, fits nicely into closets. In spite of the arbitrariness of the original choice, the business of producing primary and secondary standards of high accuracy is a very serious one which is important to business, industry, and governments. Numerous international commissions, starting in 1790, have worked together, during times of peace and war, to establish and maintain sets of standards. In the late 1800's a special alloy of platinum and iridium was cast and new sets of standard masses and lengths were made from it, compared, and sent to many different countries. The alloy that was chosen is very resistant to corrosion and can be engraved with very fine lines. In each country the standard is carefully preserved and used only for comparing secondary standards which are furnished to industry and other laboratories. Since 1960, another standard of length has been adopted in place of the platinum-iridium bar. This is the wavelength of the orange-red line in the spectrum of krypton-86. (The definition of the term "wavelength" and a discussion of the origin of the spectral lines emitted by excited atoms is given in Chapter 8.) The meter is officially defined to be a certain large number of these wavelengths.

For descriptive purposes, the meter can be visualized as a little over three

figure 3-14   The basic metric units are "human size". In every profession, however, units of convenience are chosen that are appropriate for the work being done. In printing, for instance, the unit size of type is called a "point". It is $\frac{1}{72}$ (about 0.013837) of an inch. In nuclear physics one common unit of length is the "fermi", which is $1 \times 10^{-15}$ meter. In both cases, the things being measured are small multiples of these units.

feet in length. It is divided into 100 centimeters, and 1000 millimeters. A thousand meters, the kilometer, is about five eighths of a mile.

The fundamental unit of time is, of course, the second. It is unfortunately not derived decimally, as is the metric system. The second is one-sixtieth of an hour which in turn, is one-twentyfourth of a day. The earth has too many small variations in its rotation to serve as a very accurate standard, and so new standards based on the oscillation of atomic systems have now been adopted.

*Question 3.11*    If the day were divided decimally, would one of the subdivisions be approximately the same as our second? Would a centiday be a convenient time for a lecture? Why is the second a "natural" time interval for humans?

Velocities are measured in meters per second. If the units of length and time are chosen first, there is no choice left for the units used to denote speed and velocity. It is not obvious, however, that length and time should be primary units and velocity a derived unit. So far as we know, there are no natural units of length and time. That is, there is no particular length in the universe that is more important than any other, and there is no fundamental length, like the fundamental electric charge, of which all others are multiples. There is a natural unit of velocity, however. Light always travels with the same speed, regardless of how fast the source, or the receiver, of the light is traveling. That seems to contradict common sense, but it is profoundly true. No object or information can be sent faster than the speed of light; it is a universal speed limit. Since this particular speed is so privileged, perhaps we should express all other velocities in terms of it. This is often done in physics research. The speed of light—$3 \times 10^8$ meters/sec—is called $c$, and any other speed, or velocity, $v$, is specified in terms of its $\beta$ value where $\beta \equiv v/c$. Thus the $\beta$ of an electron in a large television tube is about 0.3; it is traveling at three tenths of the speed of light.

*Question 3.12*    *About* what is your $\beta$ when you are traveling at 60 miles per hour? In converting 60 miles per hour to meters per second, use only one significant figure. (For the use of significant figures, see Appendix 2.)

The standard unit of mass was chosen by the same commissions that determined the standard meter. Called the kilogram, it is a mass of platinum-iridium that weighs about 2.2 pounds. The primary and secondary standards were compared by mass comparisons on very sensitive balances. We have defined inertial mass in terms of a momentum conservation experiment, but the same metal cylinder could serve in both definitions. There is no guarantee, however, that masses having the same inertial mass also have the same gravitational mass. Remember that we have yet to study this question. In the meantime we can use the kilogram as our standard mass. An important subdivision is $10^{-3}(.001)$ kilogram, known as the gram. One gram is almost exactly the mass of one cubic centimeter of water at $4°C$—which is one reason why the kilogram was chosen to be the size that it is. One thousand kilograms is a metric ton and (conveniently) weighs 2200 pounds, close to our standard ton.

In spite of the fact that momentum is so fundamental in many natural processes, no special name for a unit of momentum has been assigned. Momentum must be expressed as so many kilogram-meters/second—the unit of mass multiplied by that of velocity. There is, however, a special name for the unit of energy. It is called the joule and is named after James Prescott Joule, an English physicist, who in the nineteenth century determined various relationships between heat and work (see Chapter 7). The joule is a small amount of energy. If a mass of one kilogram is traveling with a velocity of one meter per second, then its kinetic energy equals $\frac{1}{2}Mv^2 = \frac{1}{2}kg(m/sec)^2 = \frac{1}{2}kg\ m^2/sec^2 = \frac{1}{2}joule$. The size of the joule is better known in terms of the unit of power—the watt. When a joule of energy is being used every second, the power expenditure is one watt. A 100 watt bulb uses 100 joules every second. The standard unit of electrical energy is the kilowatt-hour.

*Question 3.13*    How many joules in one kilowatt-hour? If a kilowatt-hour costs 4 cents, how much does one joule of electricity cost?

## summary

We have continued trying to understand the real, complex world by studying simple models. In this chapter we expanded our studies from one dimension to two. Considerable effort is needed to isolate simple phenomena from the usual complications of natural events. We made use of collisions taking place in free-fall. The simple rules of momentum and kinetic energy conservation were found to apply to two dimensions as long as we take proper account of the directional properties of momentum.

The conservation laws restrict all interactions, separating those that are possible from the larger class of those that can be imagined. Some of the effects are seen in the behavior of rockets, planets, and subatomic particles. For convenience we adopted the meter-kilogram-second system of units, and that system's defined standards for mass, length, time, velocity, and kinetic energy.

A serious restriction on the events that we have been studying so far is that they must be elastic. Kinetic energy is conserved only in those special cases where the interacting objects are not affected by denting, heating up, or rearrangement. Usually kinetic energy is lost during collisions. In the next chapter we will see how to define a new form of energy in order to preserve our law of energy conservation.

3.1   Seconds can be counted off by saying, "One thousand and one, one thousand and two, . . . ." As you can see for yourself, a reasonably dense object falls from table to floor in about half a second. If it traveled one meter horizontally in one half second, its horizontal velocity must have been about two meters per second.

3.2   Remember that the relative initial velocity, $v_o$, is determined by measuring the distance from the mark under the tee to the mark made by the bombarding marble when it is allowed to shoot off the track without any collision. If your experimental conditions are satisfactory, a head-on collision should shoot the target marble to that same mark so that $v_1 = v_o$. The bombarding marble will stop dead. Under these circumstances the conservation laws hold in a trivial fashion: $v_2 = 0$, and $v_1 = v_o$.

You will probably find that the data from the other two trials do not satisfy the momentum equation but do satisfy the energy equation.

3.3   If your experimental conditions were satisfactory, the data will satisfy the conditions stipulated in this question. Note that since the original velocity perpendicular to the forward direction was zero (necessarily!), the perpendicular part of $v_1$ must be equal in length and in the opposite direction to the perpendicular part of $v_2$.

Note the 90° condition for collision between two particles with identical mass. If the opening angle between the two paths after collision is 90°, then your experimental conditions were probably satisfactory.

3.4   The answer to this question depends upon your own analysis.

3.5   Rockets can indeed go much faster than the velocity of their exhaust gases.

The exhaust velocity has been measured *with respect to the rocket*. The more momentum is thrown out the back end, the greater the forward momentum of the remaining mass of the rocket. The rocket goes faster with respect to an observer on the earth. After a while, with respect to such an observer, the exhaust gases will be going *forward*. The situation is similar to that of a boy on the back of a train going 60 miles per hour. If the boy throws a stone backward at 30 miles per hour, it will appear to an observer on the ground that the stone is going *forward* at 30 miles per hour.

3.6   The original momentum of the shell at midtrajectory was $2Mv = 2$. This horizontal momentum must be maintained both during and after the explosion. If the horizontal momentum of half the shell is reduced to zero because it is kicked backward, the other half of the shell must carry all the momentum. Since the inertial mass of half the shell is simply $1\,M$, and its momentum must now equal $2\,Mv$, its speed must equal 2—twice the original speed at midtrajectory.

The original kinetic energy was $\frac{1}{2}(2)(1)^2 = 1$. The final kinetic energy, all carried by the half that has shot forward at twice the original speed, is $\frac{1}{2}(1)(2)^2 = 2$. Kinetic energy was not conserved; it increased.

3.7   The *change* of momentum is equal to twice the original momentum. Momentum has direction. Change of momentum = (final momentum) − (original momentum) = $(+ Mv) - (- Mv) = 2Mv$. If the final momentum was in the + ($\longrightarrow$) direction, then the original momentum was in the − ($\longleftarrow$) direction.

In order to conserve momentum, we must assume that the momentum of the wall, plus the earth to which it is fastened, changed. Since the total mass of wall and earth is so great, the recoil velocity is undetectable. If the wall were very light and mounted on rollers, you could see it recoil when the ball struck it.

3.8   Once again, if momentum is conserved there must be at least one extra invisible particle coming off in a direction opposite to that of the electron. From other evidence physicists have deduced that there are two such particles, one a neutrino, $\nu$, and the other an anti-neutrino, $\bar{\nu}$. Both are neutral, leaving no trails.

3.9   If a neutral particle decays at rest, the two decay particles must come out in opposite directions in order to conserve momentum. Instead, the two particles here have shot out in roughly the same direction, carrying a large amount of forward momentum. They must have been launched in that forward direction by the neutral particle as it decayed in flight. Note that the neutral particle on the left lost most of its momentum to the proton. It recoiled from that collision with only a small amount of momentum. Consequently, its decay particles shot away from each other in almost opposite directions.

3.10   Since the radius of curvature of a track is proportional to the momentum of the particle that made it, the two momenta can be compared by comparing the two radii of curvature. Measure them with a ruler. (It is probably easier to measure the diameters in this case—choose the diameter of the first loop.) The electron track has approximately twice the radius of curvature of the positron track, and so has approximately twice the momentum. Both lost momentum steadily as they lost energy by ionizing, and so their track curvature got smaller and smaller.

If you hold the book so that Fig. 3-11 can be seen from a glancing angle, looking down the path of the negative pi meson you can see that its track is indeed curved. The radius of curvature is much larger than that of the electron, indicating that the negative pi meson's momentum was much larger. The electron, however, was traveling much faster, as is indicated by its sparse trail. It can have a greater velocity, and *still* have a smaller momentum because its rest mass is only about $1/280^{\text{th}}$ that of the negative pi meson.

3.11   A deciday would be 2.4 hours, and a centiday would be 0.24 hours, or about 15 minutes—long enough for many lectures. The milliday would be about 90 seconds, and so the centimilliday ($10^{-5}$ day, or 10 microday) would be a little less than one second. Such a time interval is convenient or "natural" for humans because it is about the time between two normal pulse beats.

3.12   A speed of 60 mph is 88 ft/sec or about 30 meters/sec.

Thus $\beta = \dfrac{3 \times 10^1}{3 \times 10^8} = 1 \times 10^{-7}$.

3.13  One watt-second is one joule. Therefore, one kilowatt-hour equals $(1000)(3600) = (10^3)(3.6 \times 10^3) = 3.6 \times 10^6$ joules. One joule would cost only about one millionth of a cent.

### handling the phenomena

The marble collision apparatus of Fig 3-2 is reproduced as Fig. 3-15 for convenient reference. The bombarding marble rolls down the incline and collides with the target marble which rests on a "tee." Then they both fall to the floor in the same time, but meanwhile their horizontal velocities are unimpeded.

Arrange, or tape, a sheet of paper on the floor large enough to cover the region from directly under the tee to wherever the marbles will hit. Cover this with carbon paper so that the marbles will leave a mark on the paper. To avoid extra confusing marks, catch the marbles after the first bounce. Therefore, using this apparatus is usually a two person job. If you circle and number the marks after each collision, you can use the same paper for a number of trials. It is crucial to mark the point directly under the tee, the point of collision. You can do this by carefully dropping a marble held under the tee, or by dropping a plumb line from that point. After a collision you will then have three marks on the paper: the mark under the collision point, the mark where the bombarding marble hit, and the mark where the target marble hit. Draw a line from the collision point mark to the bombarding marble mark and you will have a line whose length is proportional to the horizontal velocity of the bombarding marble after the collision. Do the same for the target marble.

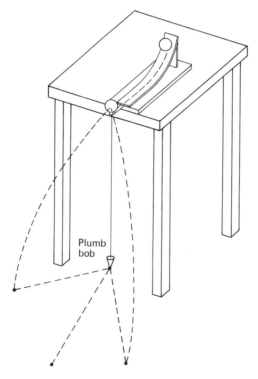

Plumb
bob

figure 3-15

There are several simple but very important precautions to take. First, make sure that the target marble is at the same height as the bombarding marble when they strike each other. If the bombarding marble "tops" (hits the top half), or "scoops" (hits the lower half), of the target marble, they will each start out with some different initial velocity in the vertical direction. Thus, they will not take the same time to fall to the floor. The second vital precaution is to make sure that the target marble is far enough away from the end of the launching track so that the bombarding marble is free of the track at the time of collision. Otherwise the collision will not be two-body because the track itself can take up some of the momentum and energy. Finally, you should make sure that for a particular experimental sequence the initial velocity of the bombarding marble is always the same. You do this simply by always releasing the marble from the same height or mark on the launching track. This initial velocity (in arbitrary units), $v_o$, is recorded on the paper by letting the bombarding marble shoot out without hitting a target. Do this several times to make sure that the marks on the paper form a small cluster, thus guaranteeing that the initial velocity remains nearly the same from one shot to the next. Draw a line on the paper from the mark directly under the tee to the center of the cluster of marks. The length of this line is proportional to the horizontal velocity, $v_o$. This line also provides the center line from which the forward and perpendicular directions of the other two lines are measured.

## derivation of a formula for the relative velocities of two objects after an elastic head-on collision. Object with mass $M_1$, and initial velocity $v_o$, strikes object, with mass $M_2$, at rest.

Conservation of momentum:

$$M_1 v_o = M_1 v_1 + M_2 v_2 \tag{3-2}$$

Conservation of Kinetic energy:

$$\tfrac{1}{2}M_1 v_o^2 = \tfrac{1}{2}M_1 v_1^2 + \tfrac{1}{2}M_2 v_2^2 \tag{3-3}$$

Simplify (3-2) by dividing both sides by $M_1$.

$$v_o = v_1 + \frac{M_2}{M_1} v_2 \tag{3-4}$$

Simplify (3-3) by dividing both sides by $\tfrac{1}{2}M_1$.

$$v_o^2 = v_1^2 + \frac{M_2}{M_1} v_2^2 \tag{3-5}$$

Square both sides of (3-4).

$$v_o^2 = \left(v_1 + \frac{M_2}{M_1} v_2\right)\left(v_1 + \frac{M_2}{M_1} v_2\right)$$

$$= v_1^2 + \left(\frac{M_2}{M_1} v_2\right) v_1 + \left(\frac{M_2}{M_1} v_2\right) v_1 + \left(\frac{M_2}{M_1}\right)^2 v_2^2$$

When simplified, this expression becomes:

$$v_o^2 = v_1^2 + 2\frac{M_2}{M_1}v_2 v_1 + \left(\frac{M_2}{M_1}\right)^2 v_2^2 \tag{3-6}$$

Subtract (3-5) from (3-6).

$$0 = \frac{M_2}{M_1}v_2^2 \left(\frac{M_2}{M_1} - 1\right) + 2\frac{M_2}{M_1}v_1 v_2 \tag{3-7}$$

Simplify (3-7).

$$0 = \left(\frac{M_2}{M_1} - 1\right)v_2 + 2v_1 \tag{3-8}$$

Solve for $\dfrac{v_1}{v_2}$.

$$\frac{v_1}{v_2} = \tfrac{1}{2}\left(1 - \frac{M_2}{M_1}\right) = \frac{M_1 - M_2}{2M_1} \tag{3-9}$$

Perform the following substitutions:

If $M_1 = M_2$

$$\frac{v_1}{v_2} = 0 \qquad v_1 = 0$$

If $M_1 = 2M_2$

$$\frac{v_1}{v_2} = \frac{M_2}{4M_2} = \frac{1}{4}$$

If $M_1 = \tfrac{1}{2}M_2$

$$\frac{v_1}{v_2} = \frac{-\tfrac{1}{2}M_2}{M_2} = -\frac{1}{2}$$

If $M_1 \gg M_2$

$$\frac{v_1}{v_2} \approx \frac{M_1}{2M_1} = \frac{1}{2}$$

proof that, in an elastic collision between two particles with equal mass, the opening angle between the paths after the collision is 90°, as shown in fig. 3-16.

Momentum conservation in forward direction:

$$M_1 v_o = (M_1 v_1 \cos\theta_1) + (M_2 v_2 \cos\theta_2) \tag{3-10}$$

Momentum conservation in perpendicular direction:

$$0 = (M_1 v_1 \sin \theta_1) - (M_2 v_2 \sin \theta_2) \tag{3-11}$$

Kinetic energy conservation:

$$\tfrac{1}{2} M_1 v_o^2 = (\tfrac{1}{2} M_1 v_1^2) + (\tfrac{1}{2} M_2 v_2^2) \tag{3-12}$$

Simplify Equations (3-10), (3-11), and (3-12) by letting $M_1 = M_2$: Under these conditions dividing Equation (3-10) by $M_1$ yields

$$v_o = (v_1 \cos \theta_1) + (v_2 \cos \theta_2) \tag{3-13}$$

Dividing Equation (3-11) by $M_1$ yields

$$0 = (v_1 \sin \theta_1) - (v_2 \sin \theta_2) \tag{3-14}$$

Dividing Equation (3-12) by $\tfrac{1}{2} M_1$ yields

$$v_o^2 = v_1^2 + v_2^2 \tag{3-15}$$

Square both sides of equation (3-13):

$$v_o^2 = (v_1^2 \cos^2 \theta_1) + (v_2^2 \cos^2 \theta_2) + (2v_1 v_2 \cos \theta_1 \cos \theta_2) \tag{3-16}$$

Square Equation (3-14)

$$0 = (v_1^2 \sin^2 \theta_1 + v_2^2 \sin^2 \theta_2) - (2v_1 v_2 \sin \theta_1 \sin \theta_2) \tag{3-17}$$

Add Equations (3-16) and (3-17) remembering that $\sin^2 \theta + \cos^2 \theta \equiv 1$:

$$v_o^2 = v_1^2 + v_2^2 + 2v_1 v_2 [(\cos \theta_1 \cos \theta_2) - (\sin \theta_1 \sin \theta_2)] \tag{3-18}$$

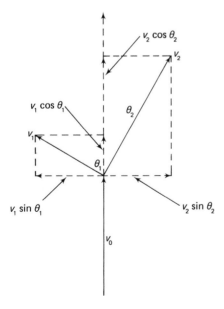

figure 3-16  Geometry of a collision between particles having equal masses. The bombarding particle had an initial velocity, $v_o$. The target particle was at rest. After collision, the bombarding particle had a new velocity, $v_1$, while the target particle had a new velocity, $v_2$. The forward and perpendicular components of velocity are labeled on the diagram.

Compare Equation (3-18) with Equation (3-15). It must be that:

$$2v_1v_2[(\cos\theta_1\,\cos\theta_2) - (\sin\theta_1\,\sin\theta_2)] \equiv 0$$

The following trigonometric identity applies here:

$$[(\cos\theta_1\,\cos\theta_2) - (\sin\theta_1\,\sin\theta_2)] = \cos(\theta_1 + \theta_2)$$

Therefore:

$$2v_1v_2[\cos(\theta_1 + \theta_2)] \equiv 0$$

Since, in general, neither $v_1$ nor $v_2$ are zero, and since the above expression must always be equal to zero for any combination of angles: $\cos(\theta_1 + \theta_2) \equiv 0$. This is true only if $(\theta_1 + \theta_2) = 90°$. Therefore the opening angle between the two paths, $(\theta_1 + \theta_2)$ must be equal to $90°$, regardless of the individual angle $\theta_1$.

A more elegant proof is geometrical. Fig. 3-17 shows the *momenta* before and after the collision. The perpendicular parts of the momenta afterwards must cancel, and the forward parts of the momenta must add up to the value of the original momentum and be in the same direction. Since $M_1 = M_2$, kinetic energy conservation requires that $v_o^2 = v_1^2 + v_2^2$. That is the Pythagorean condition for right angled triangles and can be true only if the opening angle between the two paths is $90°$.

### problems

1    A marble falling from a laboratory table top that is one meter high takes 0.45 seconds to strike the floor. It lands 80 centimeters from the base of the table. What was its horizontal velocity, in meters per second?

2    A marble is shot off a table top, striking head-on another marble held loosely on a tee. After the collision, they both hit the floor at the same time. The bombarding marble was deflected from its original path by $30°$. The target particle landed 20 cm

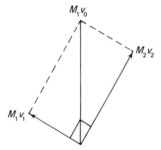

figure 3-17    The momentum geometry of a collision between particles with $M_1 = M_2$. The vector sum of the final momenta must equal the original momentum of the bombarding particle. Since kinetic energy conservation for an elastic collision of identical particles requires that $v_o^2 = v_1^2 + v_2^2$, the Pythagorean condition requires that there be a right angle between the paths taken by the two masses with velocities $v_1$ and $v_2$.

forward and 34.8 cm to one side of the original forward line of the bombarding marble. Use trigonometry, or scale diagrams, to prove that the marbles had identical mass.

3   Cannons used to be built so that the recoil would drive the cannon backward along tracks. Of course, only the horizontal part of the recoil was effective in this action; the vertical part of the recoil simply forced the cannon down against the rails. If a cannon weighed one ton and shot a ball weighing 50 pounds with a muzzle velocity of 300 miles per hour, what would be the recoil velocity if the cannon were aimed at 45° to the horizontal? Use either trigonometry or a scale diagram. Remember the conversion trick that 60 miles per hour equals 88 feet per second.

4   A skyrocket shell explodes at the height of its trajectory, breaking up into two pieces of equal mass. One is shot straight back with a speed relative to the ground equal to its forward speed just before the explosion. What is the resultant velocity of the other half? Compared with the kinetic energy just before explosion, what was the energy provided by the explosion?

5   An alpha particle (helium nucleus) has a mass approximately four times that of a proton (hydrogen nucleus). If these two particles simultaneously pass through a cloud chamber enclosed in a magnetic field, and leave trails with the same curvature, what are their relative speeds? (Assume non-relativistic speeds.)

6   The outer electrons in some atoms have speeds of about $10^6$ m/sec. What is their $\beta$? About what is the speed of a passenger jet plane in meters per second and $\beta$?

7   About what is your kinetic energy in joules when you are walking at normal speeds?

8   About how many joules of electricity per day do you use at home for lighting? (Estimate your share of the total household wattage for lights.)

*Chapter Four*

# potential energy

From observing the behavior of simple objects under very special conditions, we have deduced the existence of two conserved quantities—kinetic energy and momentum. From our everyday experience, we would not suspect that there are such conservation laws. Certainly, people did not suspect it until very recent times. Objects rolling along the ground or flying through the air appear to lose momentum constantly. Their kinetic energy fades away and soon they come to rest. The simplification necessary to find the generalization behind all the complicated facts was arrived at only after considerable effort.

Even with the simplified conditions of no friction and two-body interactions, kinetic energy can disappear. Consider what happens when the moving object runs into a spring or goes uphill. Kinetic energy disappears, but only temporarily. The moving object slows down, stops, and then speeds up again in the opposite direction, recovering all, or almost all, of the kinetic energy that it originally had. When you throw a ball straight up in the air, it has a lot of kinetic energy just after it leaves your hand. The higher the ball goes, the less kinetic energy it has, until finally, at the turn around point, it has none at all. Then it seems to gain all the kinetic energy back again as it falls.

Is it possible to define some new forms of energy so that we can maintain a conservation law? Perhaps the kinetic energy is stored in these other forms, and can be recovered from them. The energy, then, would not be lost, but merely transformed. In this chapter, we are going to define such other forms of energy in situations involving gravity, springs, and magnets. The method of defining the amount of the energy for each of these cases will be the same.

We will find out how much kinetic energy has disappeared and claim that that amount is in the other form. In that way, by definition, the energy will always be conserved. The amount of the other energy will always depend on the position of the object or on the amount of distortion of the system. The energy of position or distortion is called *potential* energy.

## 4.1 gravitational potential energy

The temporary conversion of kinetic energy is most commonly observed in the gravitational case. A roller coaster car is towed up the first and highest hill. From that point, it swoops down, gaining kinetic energy, and then shoots uphill, losing it again. To be sure, not all the kinetic energy is recovered, as we know, because the second hill must be lower than the original one. A similar situation, involving less friction, occurs in the operation of a simple pendulum like the one shown in Fig. 4-1. The pendulum bob starts moving downward from a rest position at a high point on one side of its swing. It gains speed (and kinetic energy) until finally, at the lowest point of its arc, it is traveling very rapidly. Then it swings uphill again, losing kinetic energy, until finally it comes to rest at the opposite end of its swing. The action repeats itself over and over again, with very little loss of energy to friction in the air and in the string. If we want to claim that the total energy of the pendulum never changes, then we must define the potential energy so that it is a maximum when the pendulum bob is at rest at the highest point of the swing.

*Question 4.1*   Why?

Gravitational potential energy must depend on the *height* of an object. One way to find the dependence of the potential energy on height is to measure

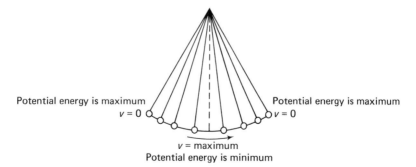

figure 4-1   A pendulum converts kinetic energy to gravitational potential energy and back again.

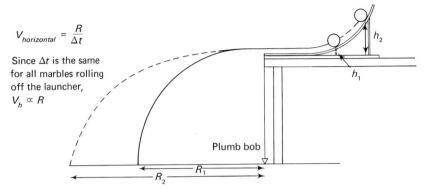

$$V_{horizontal} = \frac{R}{\Delta t}$$

Since $\Delta t$ is the same for all marbles rolling off the launcher, $V_h \propto R$

figure 4·2    Method for measuring the velocity of a marble as a function of the height through which it rolls.

the velocity of an object after it has fallen through a certain vertical distance. A simple method of doing this is shown in Fig. 4-2. Roll the marbles down the incline from progressively higher and higher positions. In each case, measure the horizontal velocity produced at the end of the launching ruler by measuring the horizontal range on the paper. Since the horizontal range is proportional to the horizontal velocity, a graph of range versus starting height is the same as a graph of velocity versus starting height. The logic of the analysis would be this: at height $h$ the marble has some unknown amount of gravitational potential energy that we want to define. After rolling down from the height $h$, this potential energy has turned into kinetic energy equal to $\frac{1}{2}Mv^2$. By measuring the velocity, we can assign a value to the kinetic energy and so define the potential energy that the marble had to begin with. The detailed directions for actually carrying out this laboratory exercise are at the end of the chapter.

*Question 4.2*    As we shall see in Chapter 6, it takes energy to make objects rotate. Since the marble is rolling and spinning as it shoots off the track, what is the effect on its velocity?

Another way to obtain the same information about gravitational potential energy is to drop the marbles from various heights and measure their velocity using the stroboscopic light technique. Then we do not have to worry about the rotational energy of the marbles. A series of strobe pictures taken of such an experiment is shown in Fig. 4-3. For each photo, the marble was dropped from the height shown. This height was measured from the finish line on which the camera was focused. The scale of the pictures is one to ten, so that a millimeter on the photo corresponds to a centimeter in the actual experiment.

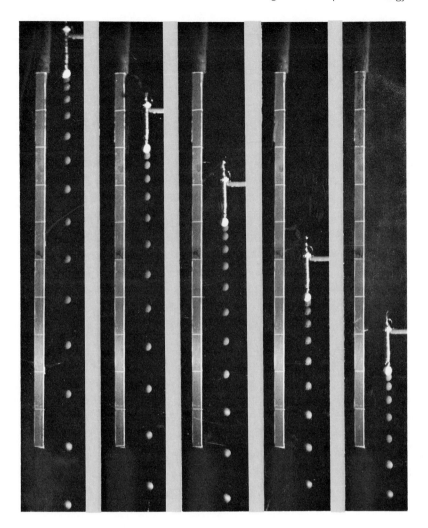

figure 4-3   In this series of strobe photos, a steel ball
bearing was dropped from five successive heights
above the zero point of the meter stick: 1, 0.8, 0.6,
0.4, and 0.2 meter. The strobe frequency was 30
flashes per second, and the scale of the printed picture is exactly $1:10$. In other words, the time between
flashes was $\frac{1}{30}$th second, and one centimeter on the
printed picture corresponds to 10 centimeters in the
original event. The approximate velocity of the ball as
it falls past the zero point can be determined by
measuring the distance it fell between flashes in that
region and dividing that distance by $\frac{1}{30}$ second. (In
some cases an average should be taken of the velocity
just above and just below the zero point.)

Since the frequency of the light flashes is given in each caption, the actual velocity of the marble as it crosses the finish line can be measured. Of course, since the marble is speeding up all the time, only an average velocity for that region can be found.

*Question 4.3*    Measure the velocity of the marble as it crosses the finish line for each of the photos. Plot a graph of the velocity (measured along the vertical axis) as a function of the dropping height (measured along the horizontal axis). Does the velocity depend upon the height in any algebraically simple way? That is, is $v$ proportional to $h$? Is $v^2$ proportional to $h$? Is $v$ proportional to $h^2$?

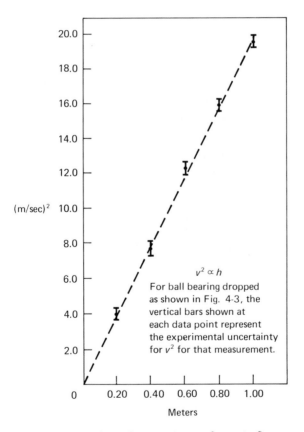

$v^2 \propto h$

For ball bearing dropped as shown in Fig. 4-3, the vertical bars shown at each data point represent the experimental uncertainty for $v^2$ for that measurement.

figure 4-4   Data from the experiment shown in figure 4-3. The final velocity squared, $v^2$, is plotted *vs* the height of the fall, $h$.

The strobe photo data are analyzed in a different way in Fig. 4-4. Here the final velocity squared, $v^2$, is plotted on the vertical axis against the falling height, $h$, on the horizontal axis. The resulting plot is a straight line, showing that $v^2$ is proportional to $h$. If the falling height is doubled, the velocity increases by a factor of four. Since the square of the velocity is proportional to the kinetic energy, then the height from which an object drops must be proportional to the kinetic energy that it obtains.

$$E_{kinetic} = \tfrac{1}{2}Mv^2 \propto h \qquad (4\text{-}1)$$

This experimental finding would be of little use if it applied only to falling marbles of that particular size. In Fig. 4-5, there is a strobe picture of many objects falling. They all started at the same time from the same height. Note

figure 4-5   Four different objects were tipped off a ledge at the same time. As they fell, three of them picked up speed at the same rate. The successive strobe flashes (0.1 second apart) caught them passing the same distance markers. The fourth object, however, did not accelerate at the same rate. It was a ball of tissue paper. Air resistance strongly affected its motion, even at low speeds, and it lagged behind the more dense objects.

that, point by point, they have all fallen the same distance, and so must be
traveling with the same velocity, picking up speed at the same rate. As we
should no longer be surprised to learn, this simple generalization is the result
of special conditions. Over 300 years ago Galileo proposed that all heavy
objects dropped from a tower would strike the ground at about the same time.
This is not what common sense would tell us, and, indeed, is not actually
the case for most falling objects. For instance, in Fig. 4-6, there is a picture
of a feather and a steel nut, dropped at the same time. The feather drifts
down much more slowly than the nut falls. In the companion picture, however,
the feather and nut fall with the same velocity. In this case, the tube was
evacuated so that air resistance was not present.

Once again we see how difficult it is to get simple conditions for experiments. Air resistance influences any falling object. The effect is immediately
noticeable with a falling piece of paper or feather. Even a falling stone,
however, will be affected when it reaches high velocity. Any object falling
freely through the air reaches a *terminal* velocity. Instead of traveling at a
steadily increasing speed, the falling object gradually attains its terminal
velocity and then continues to fall at that speed. The magnitude of the terminal
velocity depends on the shape of the falling object. For a human body it is
about 120 mph, or 60 meters per second. If the human being is wearing an
open parachute, his or her terminal velocity slows to about 15 miles per hour,
or about seven meters per second.

*Question 4.4*    Do all the objects shown in Fig. 4-5 end up with the same kinetic energy?

Since for a falling object, $v^2 \propto h$, the kinetic energy that a falling object
obtains is proportional to the height from which it falls, therefore we can

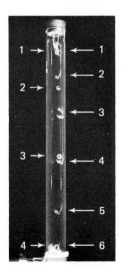

figure 4-6  Strobe photos of
feather and nut falling in
glass tubes, (left) in air, and
(right) in vacuum. (From
*Foundations of Physics* by
R. L. Lehrman and C. Swartz.
1969. Holt, Rinehart &
Winston.)

define the gravitational potential energy as proportional to $h$. How does this gravitational potential energy depend on the inertial mass, $M$, of the object? If two objects are dropped together and one has twice the inertial mass of the other, then it will at all times have twice the kinetic energy. If we claim that the kinetic energy must come from the potential energy, then the object with twice the inertial mass must originally have had twice the potential energy of the other one. In other words, gravitational potential energy must be proportional, not only to height, but also to inertial mass.

$$Mh \propto E_{grav\ pot} = ghM \qquad (4\text{-}2)$$

The proportionality constant, $g$, makes the proportionality an equation. If an object drops from a height, $h$, transforming its gravitational potential energy into kinetic energy, the situation is described by the following equation:

$$Mhg = \tfrac{1}{2}Mv^2 \qquad (4\text{-}3)$$

The proportionality constant, $g$, can be found by measuring the velocity of an object after it has fallen through a particular height.

$$g = \frac{v^2}{2h} \qquad (4\text{-}4)$$

Notice that the inertial mass, $M$, does not appear in the result. Since all objects falling from the same height and starting at rest have the same velocity, at least in a vacuum, or until air resistance dominates, then the inertial mass must cancel out of the consideration.

*Question 4.5*  Use data from the strobe photos of the falling objects in Fig. 4-3. Substitute for $v$ and $h$ and find the numerical value of the gravitational potential constant, $g$. For $v$ use units of m/sec. For $h$ use units of meters.

We have now defined a new form of energy equal to the product of the inertial mass of an object, the height to which it has been raised, and a constant derived for the particular case of an object falling near the surface of the earth.

$$E_{grav\ pot} = Mhg \qquad (4\text{-}5)$$

This expression should completely define the potential energy of an object, but does it? Take the case of a golf ball on the green of an idealized golf course. In Fig. 4-7, this situation is shown in graphical form. In these graphs, the potential energy of the golf ball is plotted along the vertical axis, and the distance along the fairway or golf green is plotted along the horizontal axis. Since the gravitational potential energy is proportional to the height,

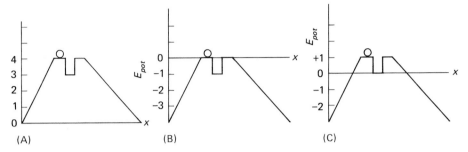

figure 4·7    Here are three different sketches of the same physical situation. The only difference from one sketch to the next is our arbitrary location of the zero point of potential energy. Since only changes in potential energy can be measured, the location of zero point does not affect calculations.

the graph looks like a profile of the hills and valleys of the golf course. In Fig. 4-7a, the golf ball is on the green at an arbitrary height above some zero point, presumably the level of the fairway. In Fig. 4-7b, the same situation is shown, but here the zero level has arbitrarily been chosen to be the level of the green itself. The golf ball, still in the same position, is now at height 0. In Fig. 4-7c, another possible but arbitrary choice of zero level has been made. Since the bottom of the cup is defined to be at 0 level, the golf ball is now at a positive height. Notice that it does not make any difference whether we label the vertical axis "height in meters", or "gravitational potential energy of golf balls in joules". Since height is proportional to the gravitational potential energy, the graphs would look the same no matter which way we labeled the axis. Nevertheless, it is surprising to find that the energy of the golf ball depends upon the arbitrary choice of zero point. Does the golf ball have a potential energy equal to $+ 4, 0$, or $+ 1$? If it drops in the cup, will its potential energy be $+ 3, -1$, or 0?

There is really no paradox in this relativistic situation with potential energy. We operationally defined potential energy in such a way that a change in it would lead to a particular change of kinetic energy. According to this definition, the height, $h$, in the case of gravitational potential energy, is not the height above something, but simply the *change* in height. In the case of Fig. 4-7a, for instance, if the ball rolls into the cup, there is a change of one unit in potential energy. The change in potential energy is equal to the final energy, $E_f$, minus the original energy, $E_o$, $(3 - 4 = -1)$. There is a negative change in potential energy because the ball has dropped. In Fig. 4-7b, the change in potential energy of the golf ball is: $E_f - E_o = [(-1) - 0] = -1$. There has been the same change in potential energy as there was in the first case. Similarly, in Fig. 4-7c, the change of potential energy of the golf ball as it drops into the cup, is equal to $(0 - 1) = -1$.

Since the only quantity that we actually measure is the *change* in potential energy, the choice of the zero point is completely arbitrary. (It is something

like the case of choosing the zero point of a temperature scale where convenience or tradition are sometimes determining factors.) For many practical
purposes, we choose the surface of the earth as the point with zero potential
energy. This means that objects above the earth will have positive potential
energy, while objects in wells or mine shafts will have negative potential
energy. We could choose the center of the earth as the zero point, but this
has no particular usefulness. There is another possibility which, at first sight,
does not appear very reasonable. That is to choose the zero of gravitational
potential energy at a point so far away from the earth that the influence of
the earth's gravity is negligible. We will see the reasonableness of this choice
in the next chapter.

*Question 4.6*   If the zero of gravitational potential energy is chosen to be a point far
away from the earth, is the potential energy on the surface of the earth
positive or negative? Why?

Let us apply the definitions of gravitational potential energy and kinetic
energy to the operation of a pendulum. In Fig. 4-1 we proposed that potential
energy at the top of the swing would turn into kinetic energy at the lowest
point. According to our definitions the total energy of the pendulum must
be equal to $Mh_1g$, where $h_1$ is the maximum height that the pendulum bob
rises from its lowest point. This *amount* of energy, (at first wholly in the form
of potential energy) stays the same, but changes so that only part of it is in
the form of potential energy. The balance of the potential energy has changed
into kinetic energy. As shown in Fig. 4-8 the pendulum bob must have only
half the potential energy at height, $h_2$, which is equal to one half of the height,
$h_1$. The other half of the energy is now in kinetic form. At the bottom of
the swing, $h_3 = 0$ and all of the original energy must now be in kinetic form.
This soon changes back into gravitational potential energy as the bob slows
down, with the potential energy reaching its maximum value at the highest
point of the pendulum's swing to the right.

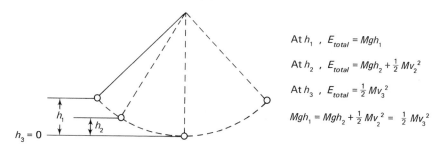

At $h_1$ , $E_{total} = Mgh_1$

At $h_2$ , $E_{total} = Mgh_2 + \frac{1}{2} Mv_2^2$

At $h_3$ , $E_{total} = \frac{1}{2} Mv_3^2$

$Mgh_1 = Mgh_2 + \frac{1}{2} Mv_2^2 = \frac{1}{2} Mv_3^2$

figure 4-8   The total energy of the pendulum remains constant at all times.

*Question 4.7*    According to this analysis, the velocity of the pendulum bob would be independent of its inertial mass. If this is true, how does the period of the pendulum depend on the inertial mass? (The period of the pendulum is the time it takes to complete one full cycle; from the highest point on the left all the way over to the highest point on the right, and back to the highest point on the left again.) Check your reasoning by making several different simple pendulums, using the same string length each time, but using bobs with different inertial masses. (See instructions under Section 2 of "Handling the Phenomena" at the end of this chapter.)

There are a number of bothersome questions about this business of defining gravitational potential energy so that it yields the appropriate amount of kinetic energy when an object drops.

For one thing, we have claimed that *gravitational* potential energy is proportional to *inertial* mass. This "explains" why all objects fall at the same rate. But, isn't this a remarkable feature of gravity? Why should energy associated with gravitation have anything to do with the inertial properties of an object? We will explore this point further in Chapter 5.

There is another problem about defining gravitational potential energy as being proportional to the height. If the definition really describes the experimental situation, then how much more kinetic energy must we provide a rocket to shoot it ten miles up compared with the kinetic energy needed to shoot it one mile up?

*Question 4.8*    What is the implication of this last question for the problem of shooting a rocket so that it will escape the earth?

## 4.2    potential energy of a spring

When a moving object is brought to rest by a spring, the kinetic energy of the object disappears and the spring is distorted. The kinetic energy can reappear, however, as the spring returns to its original shape. To conserve energy in such a case, we must find the potential energy of spring distortion so that the object's lost kinetic energy becomes the acquired potential energy of the spring. Examples of this transformation are shown in Fig. 4-9. To define the transformation quantitatively we appeal once again to strobe photos. In Fig. 4-10 the glider is shoved away by the compressed spring. The more the spring is compressed, the faster the glider shoots off. The dependence of the velocity on the amount of spring distortion can be measured directly from the pictures.

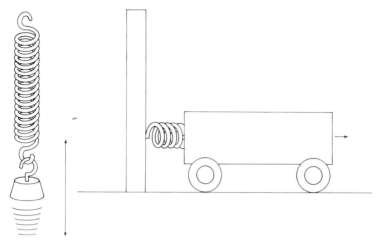

figure 4-9   Spring potential energy can change into kinetic energy and back again.

*Question 4.9*   Measure the velocity in each picture as determined by the spacing per stroboscopic flash. (Use arbitrary units with the time interval equal to one.) Plot this velocity (recorded along the vertical axis of the graph), against the compression of the spring (recorded along the horizontal axis).

In Fig. 4-11 we have plotted the velocity squared versus the compression of the spring. The plotted points do not fall on a straight line; therefore the kinetic energy produced is not proportional to the distortion of the spring. Instead, it should appear from your answer to Question 4-9, that the speed is proportional to the distortion. This must mean that the speed squared is proportional to the distortion squared. Or, in other words, the kinetic energy produced by the expanding spring is proportional to the square of its compression.

Since $v \propto x$, then $v^2 \propto x^2$ and

$$E_{kinetic} \propto x^2 \qquad (4\text{-}6)$$

Another easy way to come to this same conclusion is described in Section 3 of "Handling the Phenomena" at the end of this chapter.

It appears that springs store energy in a way that is fundamentally different from the way energy is stored in a gravitational field. For gravitation the stored energy is proportional to the vertical displacement, $h$. For springs the stored energy is proportional to the *square* of the distortion. What about the dependence of potential energy on other factors? For gravitation, the potential energy is proportional to the inertial mass of the object. Is this also true for springs?

In Fig. 4-12 we show what happens when two gliders with different inertial masses are propelled (in turn) by the same spring, compressed by the same amount. Glider #2 has twice the inertial mass of Glider #1.

*Question 4.10*    Measure the relative velocities of the two gliders and compare their kinetic energy.

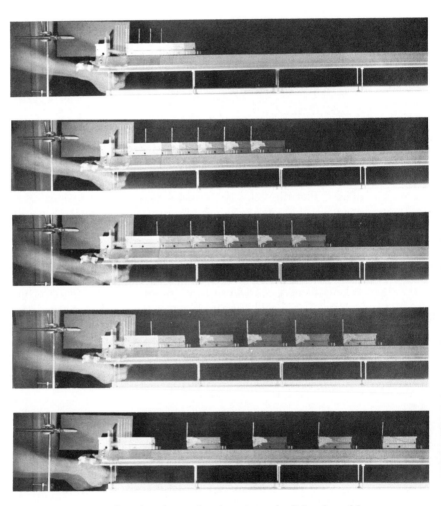

figure 4-10   Series of strobe photos showing air track glider shoved by compression spring. Velocity during time interval is proportional to spacing between vertical markers on gliders. The spring compression in the top picture is one centimeter and increases by one centimeter in each successive run.

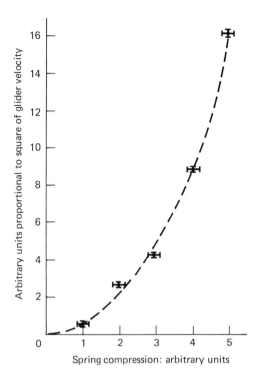

figure 4-11   Graph of $v^2$ vs compression of the spring glider $x$, for strobe shots shown in Figure 4-10.

In the case of the potential energy of springs, the stored energy is *independent* of the inertial mass of the projectile. Such a conclusion is not surprising. We would hardly expect that the energy stored in a spring would depend on the mass of a separate object with which it is going to interact. We can appeal to experience to claim that the stored energy in the spring is not related to the inertial mass of the *spring itself*. Of course, a large bronze

figure 4-12   Strobe photos of two gliders shoved by the same spring compressed by the same amount. $M_2 = M_1$. The velocity of each glider is proportional to the displacement of its vertical marker between successive strobe flashes.

spring may be able to store more energy than a small bronze spring. On the other hand, a small bronze spring can store more energy than a large lead spring. The spring potential energy depends on the nature of the spring but probably not on its inertial mass. In order to specify the dependence on the spring properties, we can write the energy equation as follows: stored spring potential energy transforms to kinetic energy

$$E_{spring \; potential} \propto x^2 \longrightarrow \tfrac{1}{2}Mv^2 \qquad (4\text{-}7)$$

$$E_{spring \; potential} = \tfrac{1}{2}kx^2 \qquad (4\text{-}8)$$

The proportionality constant, $k$, depends on the nature of the spring, and its construction, and must be determined experimentally. The factor of one half could have been absorbed into the proportionality constant, but is left separate for a reason that we will see later. We can determine the spring constant for the spring used in the photos of Fig. 4-10 by knowing the scale of the picture and the inertial mass of the glider.

*Question 4.11*    The scale of the pictures is one to twelve, the strobe light was flashing at two times per second, and the inertial mass of the glider is 0.1 kilogram. What is the numerical value of the spring constant in this case?

## 4.3   magnetic potential energy

So far we have seen two ways in which kinetic energy can be transformed into a potential form. In the case of gravity, the potential energy is proportional to the vertical displacement, and the inertial mass, of the object. In the case of a spring, the potential energy is proportional to the square of the compression, and has nothing to do with the inertial mass of the object that will gain the kinetic energy. In the first case, the stored energy has something to do with the entire earth. Since the earth is so large it is not obvious that the displacement of the object is really a separation of the earth and the object. In the second case, the potential energy is almost visibly stored in the spring. We can see that the spring is compressed. We will now consider a case where there is no visible change in the system, although the two parts of the system and their separation are evident: the potential energy between two magnets. Two different pieces of experimental apparatus are shown in Fig. 4-13. In both cases cylindrical magnets are fastened to objects that can move. Since the magnets are opposed to each other, the objects will shoot apart if they are brought close together. The sequence of strobe pictures in Fig. 4-14 shows the dependence of glider velocity on the original separation of the magnets. The task of the analysis is the same here as it was with the pictures of Fig. 4-10. We must try to find the mathematical dependence of the final velocity

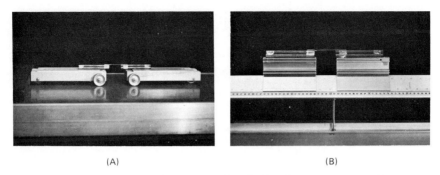

(A)                                                          (B)

figure 4-13   (A) Low friction lab carts with cylindrical magnets mounted to produce repulsion. (B) Air track gliders with cylindrical magnets mounted to
produce repulsion.

and kinetic energy on the original distortion of the system which appears here
as the original distance between the magnets.

*Question 4.12*   Measure the pictures and draw a graph showing the relative final
velocities (along the vertical axis), as a function of the original magnet
separation (along the horizontal axis). (The separation is the actual
distance between the ends of the magnets.)

Since the kinetic energy is proportional to the square of the velocity, in
Fig. 4-15, we show how $v^2$ depends on the magnet separation. Notice that
this graph looks very different from those concerned with gravitational or
spring potential energy. The magnetic potential energy, *using this particular
geometry*, is approximately proportional to the reciprocal of the magnet
separation. In symbols:

$$E_{mag\ pot} \underset{\sim}{\propto} \frac{1}{x} \tag{4-9}$$

*Question 4.13*   If this is the case, a graph of $v^2$ along the vertical axis against $1/x$ on the
horizontal axis should yield a straight line. Take your data from the
analysis of Fig. 4-14 and plot a graph to see how well this condition is
satisfied. To produce the data table from which the graph can be plotted,
set up four columns of data: (1) the speed, $v$, which in this case can be
simply the actual distance on the photograph of the displacement of the
glider between flashes; (2) the speed squared, $v^2$; (3) the separation of the
magnets, $x$; and (4) the numerical value of the reciprocal of this
separation, $1/x$.

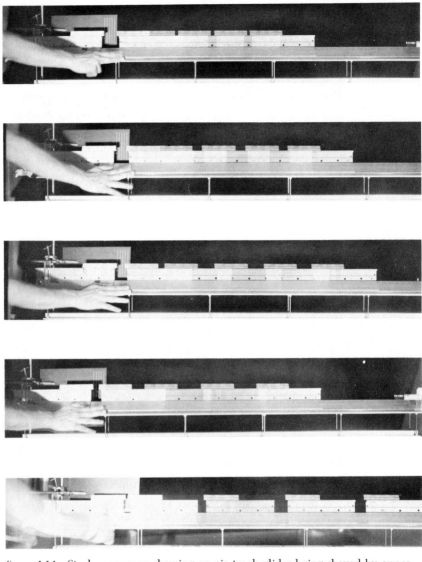

figure 4-14   Strobe sequence showing an air track glider being shoved by opposing magnets. The velocity of the glider in each picture is proportional to the displacement of either end of the glider between successive strobe flashes. In the
top picture, the "separation distance" between magnet ends was five centimeters.
In each successive picture this distance was reduced by one centimeter.

For suggestions about how you can measure the transformation of magnetic
potential energy into kinetic energy, see Section 4 of Handling the Phenomena,
at the end of the chapter.

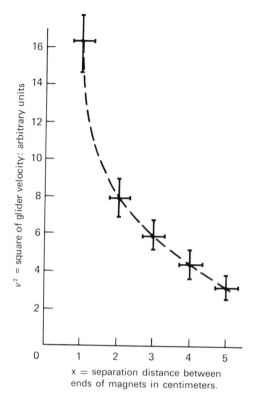

figure 4-15   The data "points" are shown as crosses that represent the uncertainty of measurement. In each case there was an uncertainty of ±0.2 in the measurement of $x$. The uncertainty of $v^2$ is larger for large values of $v$ than for small values, even though the uncertainty for $v$ is the same for all cases. Notice how different this curve is from the curve of $v^2$ vs $x$ for springs.

*y-axis label:* $v^2$ = square of glider velocity: arbitrary units

*x-axis label:* $x$ = separation distance between ends of magnets in centimeters.

## 4.4   other forms of potential energy

In three particular cases we have defined potential energy in such a way that the total of kinetic and potential energy can be conserved. In these cases we have assumed that potential energy can be converted into kinetic energy and then back again. We know that in most real cases the conversion is not 100%. Eventually the pendulum stops swinging, or a spring stops bobbing. Friction absorbs the kinetic energy and turns it into other forms of energy which we will investigate later. Nevertheless, in friction-free situations, energy can be stored and later turned completely into the kinetic form. Our method of measuring potential energy, in terms of the kinetic energy produced, guarantees that the total is a conserved quantity.

There are many other forms of potential energy. The car storage battery is an obvious example. Energy that comes from the motion of the motor is stored chemically in such a way that it can be extracted as electrical energy at a later time. The kinetic energy of the motor is provided by the stored chemical energy of the gasoline. There is energy stored in compressed air, or steam. Food can be considered a form of stored energy which animals can turn into kinetic energy. Indeed, food can be rated in terms of the number of calories it contains, and the calorie, as we shall see, is a unit of energy. Since 1945 we have all known that there is enormous energy stored within the nucleus of the atom. All of these forms of stored energy can be characterized by some kind of displacement, or rearrangement, or distortion of the parts of a system. Sometimes these are large visible parts and in other cases they are the particles and atoms of the material. In the next chapter we will explore some other features of potential energy and will apply the results to phenomena with which we are all familiar.

## summary

In this chapter we have shown a way to extend the law of the conservation of energy. We are still working with simple systems in which there is no friction or internal damage. In these systems kinetic energy disappears and yet it can be recovered at a later time. When it disappears some part of the system is displaced or distorted. We arbitrarily say that the system has potential energy, defined in such a way that the missing kinetic energy has been turned into potential energy. Three cases were explored and the potential energy measured according to the requirements of the definition. Gravitational potential energy, in the limited region that we have explored so far, is proportional to the vertical height of the displacement of an object, and is proportional to the inertial mass of the object. $E_{grav.\ pot.} = Mgh$. The spring potential energy is proportional to the square of the distortion. $E_{spring\ pot.} = \frac{1}{2}kx^2$. Magnetic potential energy is approximately proportional to the inverse first power of the magnet separation. $E_{magnetic\ pot.} \propto K/x$.

*Answers to Questions*

4.1  The potential energy must be defined so that it is a maximum when the pendulum bob is at the highest point of the swing because, at that point, the velocity of the bob and its kinetic energy are zero. The potential energy must then equal the total energy. At lower points, some of the potential energy has been turned into kinetic energy.

4.2  The gravitational potential energy due to the original position of the marble on the track is turned into the total motion energy of the marble. If this motion energy is part rotation and part motion along the horizontal direction, then the horizontal velocity must be less than it would be otherwise. It is less, for instance, than it would have been if the marble had slid down the track without friction and without rotating. In the case of our experiment, however, we will still get a *proportionality* between the release height and the resultant horizontal velocity. This is because a constant fraction of the marble's original energy is always turned into rotational energy. Since the relationship between height and horizontal velocity differs by a constant fraction because of the rolling, the type of functional relationship is not changed.

4.3   The data for this question must be obtained by measuring the strobe pictures in Fig. 4-3.

4.4   Except for the ball of tissue paper, all of the falling objects shown in Fig. 4-5 evidently have the same *velocity* at each point. Since $E_{kin} = \frac{1}{2}Mv^2$, the kinetic energy of each object must be proportional to its inertial mass. Any two of the objects have the same velocity, but the ratio of their kinetic energies must be equal to the ratio of their inertial masses.

4.5   The data for this question must be obtained from an analysis of Fig. 4-3. The result should be that $g = 9.8 \ \text{m/sec}^2$.

4.6   If the gravitational potential energy of an object is defined to be zero at a large distance from the earth, then its potential energy on the surface of the earth would be negative. It would take positive kinetic energy to raise the object to any point between the earth's surface and the zero point in which region it would have a *smaller* value of *negative* potential energy.

4.7   If the velocity of a pendulum bob is independent of its inertial mass, then the period must be independent of the mass also. That is, if two different pendulum bobs move along the same path with the same velocity, it must take them the same time. For suggestions about how to do the measurements, see section 2 of Handling the Phenomena, at the end of the chapter.

4.8   If gravitational potential energy is really proportional to the height of an object, then we would have to provide a rocket with ten times as much energy to shoot it up ten miles as we would to shoot it up 1 mile. But this means that it would take an enormous amount of energy to shoot it far away from the earth, and, indeed, an infinite amount of energy if it were to escape the earth completely. We will have to resolve this paradox in Chapter 5.

4.9   Obtain the data for the graph from the strobe pictures in Fig. 4-10. What is the relationship between velocity and distortion?

4.10   Obtain the data from the strobe pictures in Fig. 4-12. Is the spring potential energy proportional to the inertial mass of the glider?

4.11   Obtain the velocity of the glider in one of the runs in m/sec by measuring distance between positions on the strobe photo. The kinetic energy is $\frac{1}{2}Mv_1^2$ where $M$ is 0.1 kg. Equate this energy to the potential energy originally possessed by the spring, which is $\frac{1}{2}kx^2$. Solve for the value of the spring constant, $k$. (For consistent units, be sure to measure $x$ in meters. Remember the scale factor of the photo.)

4.12   Obtain the relative velocities of the glider by measuring the strobe photo in Fig. 4-14. Plot the data of $v$ vs $x$. Is $v \propto x$?

4.13   If $v^2 \propto 1/x$, a graph of $v^2$ vs $x$ is a curve with large values of $v^2$ for small $x$ and small values of $v^2$ for large $x$. The curve is one branch of a hyperbola. Plotting $v^2$ vs $1/x$, however, should yield a straight line going through the origin.

handling the phenomena

1. *Measuring Gravitational Potential Energy With Rolling Marbles*
   The apparatus for this measurement is the same marble launching device that you used in Chapter 3. In this case, no collision is required, but you will be concerned with the actual starting height of the marble.
   Make a series of runs using at least five different starting heights for the marble. Measure each height, $h$, from the starting level to the level at which the marble starts falling. For each run, measure the horizontal range by allowing the marble to strike paper, covered with carbon paper, on the floor. The range is the distance from the point directly under the end of the launcher to the mark left by the marble when it strikes. This horizontal range, $R$, is proportional to the velocity of the marble as it leaves the launcher.

$$R \propto v$$

The kinetic energy of the marble as it leaves the launcher is equal to $\frac{1}{2}Mv^2$ and therefore:

$$E_{kin} \propto v^2 \propto R^2$$

Make a table of the data as follows:

| Run Number | $h$ | $R$ | $R^2$ |
|------------|-----|-----|-------|
| 1. | | | |
| 2. | | | |
| 3. | | | |
| 4. | | | |
| 5. | | | |

Plot a graph of $h$ versus $R^2$. If you get a straight line going through the origin, then $h$ is proportional to $R^2$, which in turn is proportional to the kinetic energy of the marble after descending from the height, $h$. If $h \propto R^2$ and $E_{kin} \propto R^2$, then $h \propto E_{kin}$.
   Note the problem caused by the situation described in Question 4-2. Part of the gravitational potential energy of the marble turns into rolling kinetic energy, but the horizontal range is a measure only of the translational kinetic energy. As long as we are dealing only with rolling spheres, however, there is the same constant ratio between rolling and translational kinetic energy. Therefore, the proportionality above is still valid.

2. *Measuring the Period of a Pendulum*
   In Question 4-7 you are asked to measure the periods of pendulums with bobs of different inertial masses. It is easy to make a pendulum and hang it almost anywhere, but keep in mind the following suggestions.

For this particular exercise, use about one meter of thin flexible string (kite string will do), or stout thread. Don't use cord or rope. Tie the string to some rigid support that won't swing or vibrate. The pendulum should swing from a point such as a stationary knot. Don't let it hang from a loop that can swing back and forth with the pendulum.

The bobs can be any objects that are small compared with the length of the string, but heavy enough so that the mass of the string is comparatively negligible. A cork hung on a rope, for instance, would not be a good pendulum.

The length of the pendulum is measured from the suspension point to the *center of mass* of the bob. If the bob were a sphere, the center of mass would be at the center of the sphere. For other objects, estimate the location of the center of mass. As you try bobs of several different masses, make sure that you maintain the same pendulum length.

It is hard to make a precise measurement of one period, even with a stopwatch. Since (as you can demonstrate approximately for yourself) the period of a pendulum is almost independent of the amplitude of swing, you can time five or ten periods and take an average. Your precision will then be considerably better.

3. *Another Way of Measuring the Potential Energy of a Spring*
To measure the potential energy of a system, cause it to transform into the kinetic energy of some object. Then measure the inertial mass and velocity of that object and the relation:

$$E_{potential} \longrightarrow E_{kinetic} = \tfrac{1}{2}Mv^2$$

will hold.

If the object is subject to friction and slows down rapidly, the velocity must be measured right after the potential energy has been turned into kinetic energy. That is why strobe photos were used to record the velocity of the falling marbles and the propelled air track gliders. An alternative in the case of marbles was to let them roll down a curved track, converting vertical fall to horizontal motion. Once released from the track, the marbles could travel almost without friction. Since the time of fall from track to floor remains constant, the horizontal range is a relative measure of the horizontal velocity, which is the original velocity given to the marble from the conversion of gravitational potential energy.

With a slight variation, (Fig 4-16), the same type of free fall apparatus can be used to measure the potential energy of springs as a function of spring distortion. In this case, the track should be flat along the table. Mount a compression spring so that it will shove horizontally against the sphere on the track. In order not to complicate the situation, the spring should be much less massive than the ball. Instead of marbles, use heavy steel ball bearings or croquet balls. The spring should be small and light. Make measurements of horizontal range (which is proportional to velocity) for at least five different spring compression distances. Plot the results on a graph and determine the relationship between

$$E_{spring\ pot.}\ \text{and}\ x.$$

If you have an air track available, the measurement of $v$ versus $x$ can be made with fair precision without a strobe light. Simply use a stopwatch to time the flight of the propelled glider along one length of the track. The velocity is inversely proportional to the flight time: $v \propto 1/t$. Care must be taken to make sure that the

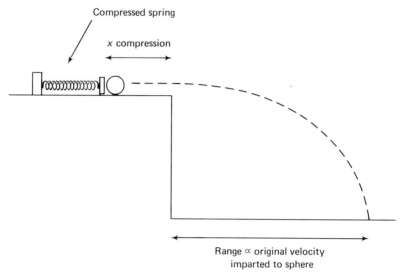

figure 4-16

glider does not scrape or bump during its release, and while it is being shoved by the spring.

4.    *Other Ways of Measuring the Potential Energy of a Magnet System*

The pictures in Fig. 4-14 are strobe photos of an air track glider being shoved by opposing magnets. The same experiment can be done, with less precision, simply by timing the flights of the glider along one length of the track. Note the suggestions and precautions given in the previous section.

The experiment can also be done by modifying the apparatus that shoots a marble off a table as shown in Fig. 4-17. In this case the marble should be mounted on a small car that is free to roll toward the edge of the table. The opposing magnet

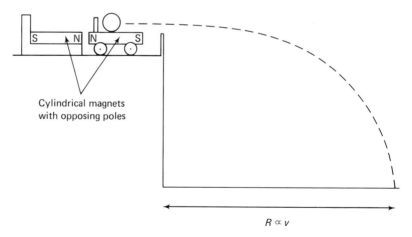

figure 4-17

can be fastened to the car, while the marble should rest on the car in such a way that it can roll off easily when the car hits a stop at the edge of the table. The other magnet should be permanently fastened to the table at the same height as the magnet mounted on the car. Hold the car so that the opposing magnets are separated by a distance, $x$. Release the car, which will roll to the edge of the table. When it hits the stop, traveling with velocity $v$, the marble will shoot off, and fall to the floor, with a horizontal range, $R$. Then, $R \propto v$. Measure $R$ for at least five different separation distances, $x$.

Note one important feature of the conversion of magnetic potential energy to kinetic energy. The process does not depend on the inertial mass of the magnets. Gravitation is the only form of potential energy that is proportional to the inertial mass of the object having the potential energy.

## problems

1   A crossbow shoots a bolt straight up with an initial velocity of 60 meters per second. (*About* how fast is this in miles per hour?) Neglect the energy lost to air friction and assume that the original kinetic energy is turned completely to gravitational potential energy. If the bolt has a mass of 200 grams, how high will it rise? How high will it rise if it has twice this mass?

2   A roller coaster car loses ten percent of its initial gravitational potential energy to friction just going down the first slope. As it climbs the second slope, friction uses up ten percent of the kinetic energy that it had in the valley. If the height of the first hill is 20 meters above the valley, what is the velocity of the car when it reaches the valley, and what is the maximum height of the second hill?

3   A spring gun shoots a marble straight up to a height of 30 cm after the spring is compressed 1 cm. What will the initial (muzzle) velocity of a marble be and how high will the marble rise if the spring is compressed 3 cm?

4   Choose the zero of gravitational potential energy to be at the surface of the earth. In that case, the potential energy of an object in a well is negative. Suppose a one kilogram rock drops from a point five meters down a well to the bottom, which is 30 meters beneath the surface. What is the final potential energy when the rock hits the bottom of the well? What is the initial potential energy when the rock is five meters down? What is the change in potential energy, and is this change positive or negative? What is the final velocity just before the rock hits?

5   There is a light spring between two air track gliders, one with a mass of 200 grams and the other with a mass of 400 grams. If the spring has a constant of 100 newtons per meter and is initially compressed two centimeters, what are the velocities of the gliders when they first spring apart? (First, remember that momentum must be conserved: this requirement determines the *ratio* of velocities of the two gliders.)

6   The electrostatic potential energy, $E$, between two point (or spherical) electric charge distributions is proportional to $1/r$, where $r$ is the distance between the centers of the charges. Sketch a graph of this dependence of $E$ on $r$ from $r = 0$ to $r = 5$. (Graph $E_{elec.\ pot.} = K/r$, with the horizontal axis units for $r$ chosen to be 1, 2, 3, etc. Choose the vertical axis units for $E$ so that $E$ goes from 0 to 4 with $K = 1$.)

7   Two cylindrical magnets are placed in a smooth trough so that they will be pushed

away from each other. Their magnetic potential energy is $E_{mag.\ pot.} = 16/x$, where the separation distance, $x$, is in centimeters. They are prevented from separating too far by a light bronze spring which is arranged so that it is in its equilibrium, unstretched, condition when the magnets are touching—that is, when $x = 0$. The spring potential energy is: $E_{spring\ pot.} = x^2$, where $x$ is in centimeters. Evidentally, the lowest magnetic potential energy is obtained by letting the magnets separate as far as possible. On the other hand, the lowest spring potential energy is obtained when the magnets are touching and the spring is unstretched.

On the same graph, plot the magnetic and spring potential energy. The potential energy of the system is the sum of these two. Add the two curves, point by point, to obtain the combined potential energy curve. The magnets will come to equilibrium at a separation distance that makes the total potential energy a minimum. Use your graph to determine this distance.

8    Use pendulums of several different lengths to measure the dependence of period on length. Since you have already measured the period for a length of one meter, try half a meter. Is the period one half or double that of the one meter pendulum—or some other value? Try a very short length, such as one fifth or one tenth of a meter. What is the relationship between period and length?

# forces

In Chapter 4 we defined three kinds of potential energy—due to gravity, cylindrical magnets, and springs. Potential energy was defined so that a change in potential energy would produce an equal change in kinetic energy. In Chapter 6 we will consider cases where not all of the potential energy is transformed into kinetic energy. We can define new types of energy in order to preserve the conservation law, but the methods of measurement will become more complicated. Before getting into those complications, we can use our simple friction-free examples to define another important quantity.

## 5.1   the definition of force

Take another look at our special golf course as shown in Fig. 5-1. Golf balls are shown at three different positions, all at the same height and, therefore, all having the same gravitational potential energy. Their actual physical situations are quite different, however. The golf ball in the hole is motionless and will remain so. The ball on the slope will roll downhill gathering speed as it goes, and the one over the edge of the green will plunge straight down. Evidently the *graph* of potential energy, as a function of the distance through which the object can move, contains much more information than simply the value of the potential energy of the object at one point. The single value of potential energy contains no information about the stability of the object.

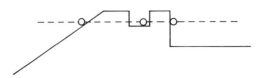

figure 5-1    Schematic cross section of a golf green. The ball is at the same height and therefore has the same potential energy in all three positions. However, the future behavior of the ball is radically different in the three cases.

For such information, we have to know the slope of the potential energy curve. If the slope is zero, or in other words if the potential energy curve is horizontal, the object will stay in place. A positive slope, going uphill to the right, means that the object will roll downhill to the left. A negative slope, going downhill to the right, will send the object rolling to the right. The steeper the slope, the greater the tendency of the object to change its position.

Slope is defined as the ratio of "rise" to "run". On a graph, this is the ratio of the change in vertical height to the change in horizontal position. In each case, the change must be measured in the units specified by that particular axis. Thus in Fig. 5-2, where $E$ is potential energy, and $x$ distance, the slope at $A$ is equal to $\Delta E/\Delta x = 1$ joule/1 meter, where $\Delta$ means a small change in the quantity. At position $B$ of Fig. 5-2, the slope is 0 because $\Delta E = 0$. As $x$ increases there is no change in $E$. At point $C$, the slope equals $-1$ joule/$+2$ meters $= -\frac{1}{2}$ joule per meter. The slope is negative because an increase in $x$ leads to a decrease in $E$.

The slope on the graph is a measure of the tendency of an object to change its position. In everyday language we describe this tendency in terms of the *force* acting on an object. We can arbitrarily define force as being equal to the negative slope of a graph of potential energy as a function of displacement: $F = -(\Delta E/\Delta x)$. The negative sign indicates the proper direction of the force.

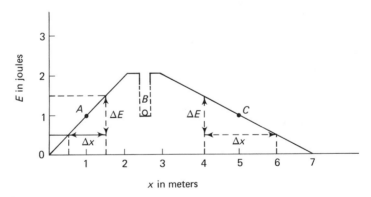

figure 5-2    The slope of the golf green, $\Delta E/\Delta x$, determines the future behavior of the ball at each point.

For instance, in Fig. 5-2 the slope at $A$ is positive. According to our definition, the force would be negative; which means, directed toward small $x$. That is just the direction we would expect for the force; pointed toward the left. At point $C$ on that graph the slope is negative. The force is therefore defined as being positive; pointed to the right, in the direction of larger $x$.

This particular definition of force is unusual because it presents force as a derived, rather than as a primary, quantity. In this text we have emphasized the conservation laws, particularly the conservation of energy. Instead of starting out by defining a force as a "push or pull," we have analyzed the nature of energy and now define force in terms of energy. This approach reflects the realities of modern atomic and nuclear research, where energy levels usually have more fundamental significance than any kind of force. This approach also has surprising power in some classical descriptions of objects and their behavior.

The effective force acting on an object not only depends on where the object is, but also on the direction in which the object can move. Unless restrained by some other mechanism an object will always try to lose its potential energy by moving to a region of smaller potential energy. However, if the object is at rest, it will start to move in such a way that the potential energy will be reduced with the smallest displacement of the object. In other words, it will start to move in the direction of the greatest force. Consider, for example, a ball on the side of a hill. It could reduce its potential energy by rolling down slowly at a small angle. However, when it is released, unless it is forced to go that way by a track, the ball will start to roll on the steepest path downhill.

Since we will want to measure forces and use them quantitatively, we must assign a standard unit to the quantity. From our definition, it appears that the unit of force is equal to one joule per meter. This is called, appropriately enough, one newton. Newton's laws deal with forces. He was the first to define them and their effects in a quantitative, and rigorous way. Although we will not cite these laws by number or with the usual derivations, in this chapter we will present their content from a different viewpoint.

Weight is simply the gravitational force on an object. Since it acts in a vertical direction its magnitude can be determined from the graph shown in Fig. 5-3. The gravitational potential energy as a function of vertical height yields our familiar straight line. The algebraic description is: $E_{grav\ pot.} = Mgh$. The vertical downward force (in other words, the weight) is simply equal to the negative of the slope of that curve. It is a constant slope equal to:

$$\frac{\Delta E}{\Delta h} = \frac{\Delta(Mgh)}{\Delta h} = \frac{Mg\Delta h}{\Delta h} = Mg.$$

An extremely important point is hidden in this simple expression. Note that the *weight* of an object is proportional to its *inertial mass*. Later in this chapter we will consider the implications of this fact.

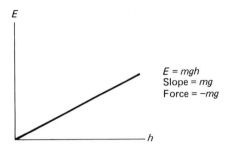

E = mgh
Slope = mg
Force = −mg

figure 5-3   Graph of gravitational potential energy as a function of height, *h*.

As we saw in Chapter 4, the numerical value of the proportionality constant, *g*, is equal to 9.8. In our units, it equals 9.8 newtons per kilogram. One kilogram, therefore, *weighs* 9.8 newtons.

*Question 5.1*   (a) In the English system, a kilogram weighs 2.2 pounds. What is your approximate weight in newtons? (b) Since force is defined as being equal to the negative slope of the potential energy versus displacement curve, is the weight, according to this definition, in the right direction?

## 5.2   the force exerted by a spring

In Chapter 4 we operationally defined the potential energy of a spring by turning the potential energy into the kinetic energy of an object. This kinetic energy was then measured. The spring potential energy was found to be proportional to the square of the distortion: $E_{spring\ pot.} \propto x^2$. The proportionality constant does not depend on the mass of the spring, but does depend on the material from which it is made and the way it is wound. Each spring can be described by a proportionality constant so that: $E_{spring\ pot.} = \frac{1}{2}kx^2$. The curve showing this dependence of the potential energy, *E*, on the amount of distortion, *x*, is reproduced in Fig. 5-4. The restoring force exerted by the distorted spring is defined as being equal to the negative slope: $F = -(\Delta E/\Delta x)$.

*Question 5.2*   Does the definition of force qualitatively match the force that you expect to find exerted by a spring, both in direction and magnitude? For instance, is the force large or small for a small distortion? If you increase the distortion of a spring toward the right, what is the direction of the force?

Slope lines are carefully drawn at a number of points in Fig. 5-4. Directly underneath is another chart which can be used to plot the force exerted by the spring as a function of its distortion. For instance, the slope at point $B$ is equal to $\Delta E/\Delta x = 10$ joules$/1.7$ meters $= 5.9$ newtons. That particular value is plotted in the lower graph as the magnitude of the force at the position $B$. Since the slope is positive, the force is negative, implying that it is toward the left, tending to decrease the distortion.

*Question 5.3*  From the slopes shown on the other points of the curve in Fig. 5-4, compute the force at each point and plot the rest of the curve of force on the graph.

The force exerted by a spring can also be found algebraically from the definition $F = -(\Delta E/\Delta x)$. The change in potential energy, $\Delta E$, is the difference between the final energy, $E_f$, and the original energy, $E_o$. The change in distance, $\Delta x$, is the difference between the final position, $x_f$, and the original position, $x_o$, of the object.

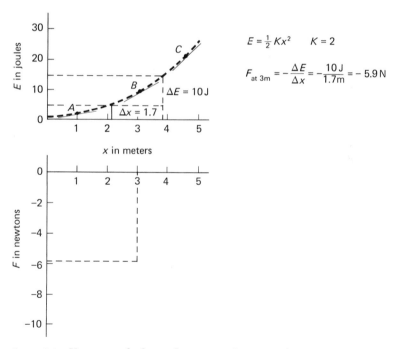

$$E = \tfrac{1}{2} K x^2 \qquad K = 2$$

$$F_{\text{at 3m}} = -\frac{\Delta E}{\Delta x} = -\frac{10\,\text{J}}{1.7\,\text{m}} = -5.9\,\text{N}$$

figure 5-4  Upper graph shows the potential energy of a spring as a function of its distortion, $x$. (Below.) A graph of the force exerted by the spring can be generated from measurements of slopes at various points of the potential energy curve.

$$F = -\left(\frac{E_f - E_o}{x_f - x_o}\right)$$

$$= -\left(\frac{\frac{1}{2}k(x + \Delta x)^2 - \frac{1}{2}kx^2}{\Delta x}\right)$$

$$= -\left(\frac{\frac{1}{2}k(x^2 + 2x\Delta x + \Delta x^2) - \frac{1}{2}kx^2}{\Delta x}\right)$$

$$= -\left(\frac{\frac{1}{2}kx^2 + kx\Delta x + \frac{1}{2}k\Delta x^2 - \frac{1}{2}kx^2}{\Delta x}\right)$$

$$= (-kx - \tfrac{1}{2}k\Delta x) \longrightarrow -kx \text{ for small } \Delta x.$$

The quantity, $(\frac{1}{2}k\Delta x)$, is negligible compared with $(kx)$, especially as $\Delta x$ gets smaller and smaller, which is the case as the exact slope of the line at a particular point is taken. *Therefore, the force exerted by a spring is proportional to the amount of distortion of the spring.* The negative sign says that the force acts in the direction opposite to the distortion. If the spring is pulled to the right, the force is to the left. Now we can see the reason why the original expression for the spring potential energy was written as $\frac{1}{2}kx^2$. In the expression for the force exerted by the spring, the factor of $\frac{1}{2}$ disappears, leaving the formula as, simply, $F = -kx$.

Most ordinary springs follow this rule, called Hooke's Law, unless they are stretched too far. The proportionality between the restoring force of a spring and the length of its distortion makes this a very convenient way to measure force. There are standard spring balances, such as those found in ordinary bathroom scales, calibrated in either newtons or the English unit of pounds. Typical scales are shown in Fig. 5-5. The one on the left is exerting a horizontal force against the hook on the wall. The one on the right is exerting a vertical force.

*Question 5.4*    The spring force in Fig. 5-5b, is directed upward in a direction that decreases the amount of spring potential energy. Why does the object not move upwards so that this will happen?

In Section 1 of "Handling the Phenomena" at the end of this chapter, you can find directions for checking Hooke's Law experimentally.

## 5.3    the force between cylindrical magnets

In Chapter 4 we converted magnetic potential energy into the kinetic energy of friction free gliders. The magnetic potential energy defined in this way was approximately inversely proportional to the magnet separation:

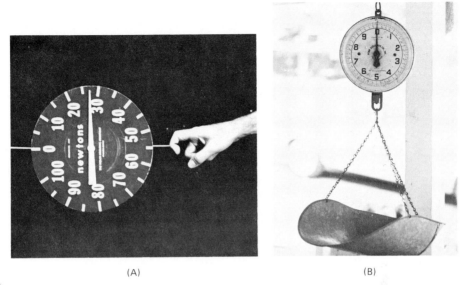

(A)                                                                                    (B)

figure 5-5   (A) Laboratory spring scale, calibrated in newtons. (B) Spring balance
at a roadside farm produce stand, calibrated in pounds.

$E_{mag.\ pot.} \propto 1/x$. The data from a similar experiment is presented in Fig. 5-6.
Slope lines are drawn at various points on the curve. At point $B$, the slope

equals $\Delta E/\Delta x = \dfrac{(-0.14)}{0.014} = 10\text{N}$. That value of force is plotted at the appro-

priate position on the force curve directly underneath.

Question 5.5   (a) Are the slopes at the various points on the energy curve qualitatively
what you would expect in predicting the force between the two repelling
magnets? That is, does a small separation produce a large force or a small
force? If the force is in the direction that produces a larger separation,
should it be positive or negative? (b) Evaluate the other slopes shown on
the energy curve and plot the corresponding curve of force as a function
of separation distance.

The actual curve, of the energy graph of two magnets as a function of
their separation distance, is not exactly $1/x$ because there is a small difficulty
about defining the distance between the two magnets. In fact, if the distance
is taken as the distance between the *centers* of the two magnets the potential
energy should be proportional to $1/x^2$. However, defining the separation
distance as the distance between the ends of the magnets, the potential energy,

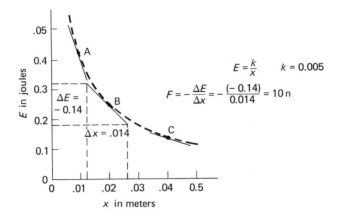

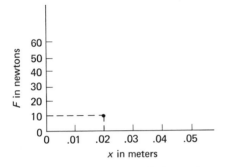

figure 5-6   The upper graph shows the potential energy between two cylindrical magnets as a function of the separation distance, $x$, between their ends. (Below.) A curve of the force between the two magnets can be generated from measurements of the slopes at various points of the potential energy curve.

as we saw experimentally, is approximately equal to $k/x$. We can algebraically deduce the force between the magnets by appealing to the definition of force.

$$F = -\frac{E_f - E_o}{x_f - x_o}$$

$$= -k\frac{\left(\dfrac{1}{x + \Delta x}\right) - \left(\dfrac{1}{x}\right)}{\Delta x}$$

Multiplying $1/(x + \Delta x)$ by the quantity, $x/x$, (equal to one); and $1/x$ by the quantity, $(x + \Delta x)/(x + \Delta x)$, (also equal to one) allows these two fractions to be expressed with a common denominator

$$-k\frac{\left(\dfrac{x}{x(x+\Delta x)}\right)-\left(\dfrac{(x+\Delta x)}{(x+\Delta x)x}\right)}{\Delta x}$$

The common denominator, $(x + \Delta x)x$, now allows us to make the following simplifications

$$= -k\frac{\dfrac{x-(x+\Delta x)}{(x+\Delta x)x}}{\Delta x}$$

$$= -k\frac{-\Delta x}{(x+\Delta x)x\Delta x}$$

$$= k\frac{1}{(x+\Delta x)x}$$

$$= k\frac{1}{x^2+x\Delta x} \approx k\frac{1}{x^2} \text{ when } \Delta x \text{ is small.}$$

For the same reasons given before, the quantity $x\Delta x$ is negligible compared with the quantity $x^2$. What we have found is that in a system where the potential energy is proportional to $1/x$, the force at any point is proportional to $1/x^2$. Notice how this relationship describes the curve that you plotted in Fig. 5-6. The potential energy of a system containing two magnets increases rapidly as the distance between them decreases, but the force tending to shove them apart increases even more rapidly. You can demonstrate this for yourself by following the directions in Section 2 of Handling the Phenomena at the end of this chapter.

*Question 5.6*   If the two magnets were aligned to attract each other, what would the graphs of potential energy and force look like?

## 5.4   the relationship of work and energy

In our development we started out by defining the conserved quantity called energy. Force is a derived quantity defined (so far) only in situations where there is a potential energy as a function of position or distortion. The logic can be turned around in order to answer the question, "What happens when a distorted system exerts a force to move an object through a distance?" Since

$F = -(\Delta E / \Delta x)$, then $F \Delta x = -\Delta E$. If a system reduces its distortion by exerting a force, through a distance, then the reduction of potential energy is equal to $F \Delta x$. This product of force and displacement is given the technical name of *work*. If a system, such as a spring, does work by pushing some outside object, then the spring energy decreases. If an external force compresses the spring through a distance $\Delta x$, then the spring stores up energy equal to $+\Delta E$.

The word, work, is used in many different ways in ordinary language. The common uses agree with the technical use in so far as everyone assumes that it takes energy to do work. Notice, however, that simply exerting a force does not necessarily require that work be done or energy be used. For instance, a C-clamp can exert a continual force holding two boards together without using any fuel. To use up fuel, and so to do work, the force must be exerted through a distance. A special case for the relationship between force and work is shown in Fig. 5-7. That is the case of a constant force being exerted against gravity. The force is plotted along the vertical axis, and the upward displacement along the horizontal axis. In order to lift a one kilogram mass, an upward force of 9.8 newtons (just equal to its weight) is necessary. If that much force is exerted on the one kilogram mass, then it can be raised at a constant velocity. The graph shows a constant force of 9.8 newtons exerted over a vertical displacement of two meters. According to our definition, the work done is:

$$F \Delta x = (9.8 \text{ newtons})(2.0 \text{ meters}) = 19.6 \text{ joules}$$

On the graph, this corresponds to the rectangular area underneath the force curve. After someone has done 19.6 joules of work on the object, it has more gravitational potential energy. That energy has been increased by $Mgh$, where $h$ is the vertical distance moved. Numerically, the increase in potential energy equals the work done on the object, since $M = 1$ kilogram, $g = 9.8$ newtons per kilogram, and $h = 2$ meters.

*Question 5.7*    Why is it not necessary to exert a constant force greater than the weight in order to lift an object?

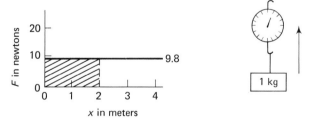

figure 5-7    A constant force of 9.8 newtons lifts a weight two meters. The product of the force used and the distance through which the force, $w$, was exerted, is represented on the graph by the crosshatched area.

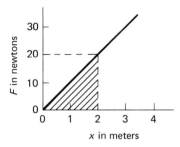

figure 5-8  A variable force is exerted through a distance. In this case the force was proportional to the displacement: $F = kx$. Once again, the area under the curve represents the work done.

A slightly more complicated way of doing work on a system is shown in Fig. 5-8. If work is done in stretching a spring, the force necessary depends on the amount of the stretch, $F = kx$. This relationship is shown in the graph. The work done is equal to some average force times the change in the length of the spring. *Since $F \propto x$, the average force is simply one half of the final force.* Therefore, the work needed to stretch the spring to a distance $x$, is equal to one half the final force, $F_f$, times $x$: work $= \frac{1}{2}F_f x$. Since $F_f = kx$, work $= \frac{1}{2}(kx)x = \frac{1}{2}kx^2$. The work done in stretching the spring a distance $x$, is just equal to the increase in the spring potential energy. On the graph this work is represented by the triangular area under the force curve. Since the area of a triangle is equal to one half the base times the height, the graphical solution to the problem of work done is equal to the algebraic solution.

*Question 5.8*   Explain in words the significance of the factor of $\frac{1}{2}$ in the expression for potential energy of a spring.

Force has been defined in terms of a change in the potential energy of a system. What happens if a force is exerted on an object for some distance, but there is no change in potential energy? Suppose that a force is exerted in the direction of motion on a friction free glider while it moves over a distance of one meter. The force could be provided by a stretched spring. As long as the spring is stretched the same amount, it must exert a constant force on the glider. Since the glider remains at the same height from the earth, there is no change in the gravitational potential energy. There is a change, however, in the kinetic energy of the glider. The force will make the glider go faster and faster. Similarly, if a force larger than the weight of an object is exerted on it in the upward direction, it will not only lift the object, increasing its potential energy, but will also provide it with kinetic energy. This fact should not be surprising since doing work on a system increases its energy, and that energy can appear in either kinetic or potential form. It can also appear internally, in forms related to the structure of the material, but we will investigate that situation in coming chapters.

Since work can produce a change in kinetic energy, we can define the force exerted by an object in terms of the change in its kinetic energy over a distance. In symbols, $F = -(\Delta(\frac{1}{2}Mv^2)/\Delta x)$. Assuming that there is no change in the inertial mass of the object, a change in its kinetic energy must be caused by a change in its velocity. After traveling a short distance, $\Delta x$, the final velocity, $v_f$, must be equal to $v_o + \Delta v$. Then the change in kinetic energy equals the final kinetic energy minus the original kinetic energy.

$$\frac{\Delta E}{\Delta x} = \frac{\frac{1}{2}Mv_f^2 - \frac{1}{2}Mv_o^2}{\Delta x}$$

$$= \frac{\frac{1}{2}M(v_o + \Delta v)^2 - \frac{1}{2}Mv_o^2}{\Delta x}$$

$$= \frac{\frac{1}{2}M(v_o^2 + 2v_o\Delta v + \Delta v^2) - \frac{1}{2}Mv_o^2}{\Delta x}$$

$$= \frac{Mv_o\Delta v + \frac{1}{2}M\Delta v^2}{\Delta x} \longrightarrow \frac{Mv_o\Delta v}{\Delta x} \text{ for small } \Delta v.$$

As usual, the small term containing $\Delta v^2$ is negligible (compared with the other term) for very small changes in velocity. Now, since velocity is equal to the displacement, $\Delta x$, divided by the time, $\Delta t$, taken for the displacement, then: $v_o = \Delta x/\Delta t$. Therefore:

$$\frac{\Delta E}{\Delta x} = Mv_o\frac{\Delta v}{\Delta x} = M\frac{\Delta x}{\Delta t}\frac{\Delta v}{\Delta x} = M\frac{\Delta v}{\Delta t}$$

Assuming that the inertial mass, $M$, does not change, $M\Delta v$ is just the change of momentum. It appears that the ratio of the change in the kinetic energy of an object over a small *distance* is equal to the ratio of the change in *momentum* of the object divided by the short *time* in which it takes place. This relationship provides us with another definition of force, useful for cases where an object suffers a change in its motion but not in its potential energy.

$$F = -\frac{\Delta E}{\Delta x} = -\frac{\Delta(momentum)}{\Delta t}$$

*Question 5.9*  If a one kilogram ball strikes the floor traveling at one meter per second and bounces off with the same velocity upwards, what is the average force exerted on the floor if the time of contact between the ball and the floor is 0.1 second? According to the definition, what is the direction of the force exerted on the floor by the ball?

The average force exerted by an object changing its momentum is vitally dependent upon the time in which it happens. If you are traveling at 60 mph in a car, you suffer the same change of momentum whether you come to rest over a time lasting 10 seconds or thousandths of a second. However, in the latter case, the average force exerted on your seatbelt (or on the windshield) is 1,000 times as great.

## 5.5   another look at gravitational potential energy

We have examined three kinds of potential energy—gravitational, magnetic, and that from a spring. It also appears that some of the same treatment can apply to kinetic energy; since, in a very real sense, an object in motion has the potential to do work. In our preliminary treatment of gravitational potential energy we passed over two important points.

1   Is it not surprising that gravitational potential energy is proportional to inertial mass? Inertial mass plays no such role in any of the other forms of potential energy. It is because gravitational potential energy is proportional to inertial mass that all objects fall at the same rate under conditions where air resistance is negligible. If an object drops from a height, turning its gravitational potential energy into kinetic energy, then:

$$Mgh = \tfrac{1}{2}Mv^2$$

The inertial mass cancels out of the equation, leaving the velocity of a freely falling object independent of its inertial mass.

Since the gravitational potential energy of an object is proportional to its inertial mass, then its weight is also proportional to its mass. This situation provides us with a new way of measuring the inertial mass. Our original defining measurement for inertial mass consisted of a dynamic experiment. An object with an unknown inertial mass was repelled from an object with unit inertial mass. Since momentum, by definition, is conserved, the unknown inertial mass, $M_2$, is equal to the ratio of the velocity, $v$, of the unit mass, $M_1$, to the velocity, $V$ of the unknown inertial mass. Thus:

$$M_2V = M_1v \text{ which simplifies to } M_2V = 1v$$

Therefore;

$$M_2 = \frac{v}{V}.$$

Since the weight of an object is proportional to its inertial mass, we can also determine its inertial mass with a static experiment. Two examples are shown in Fig. 5-9. One of these is the familiar pan balance. It compares the weight of an unknown object to the weight of standard units in the other pan. In the other case, the weight of the unknown object is balanced by the upward force exerted by the extended spring. The scale reads the amount of this extension in terms of forces that were derived from a previous calibration of the spring's extensions due to various standard masses.

figure 5-9    The gravitational force can be measured
in a static situation with a pan balance or with a
spring balance. In the first case an unknown weight
is balanced against standard weights. In the second
case, the unknown weight is balanced by a standard
or calibrated spring.

Notice how entirely different these two types of measurement are. With our
original method, the objects were shot into motion and the mass depended on
the reluctance of the object to change its motion. In the second type of measurement, nothing is moving. An entirely different property of the mass appears. Mass
serves as the source of attraction between the object and the earth. Why should
the reluctance of an object to change its velocity have anything to do with the
gravitational attraction that it exerts on another object? See Fig. 5-10. We usually
think of mass in terms of its gravitational properties—massive, usually means
heavy. The symbol for gravitational mass is $m$, and what we have observed is
that $m \propto M$. With the appropriate choice of units (kilograms in both cases) we
can set $m = M$. Determining this equality was really the whole point of the
experiment, described by Galileo, of dropping stones from a tower. Objects fall
at the same rate only if $m \propto M$. That experiment of dropping stones of different
masses through air cannot give very precise results. Far more sophisticated
experiments were done in the 1890's by the Hungarian physicist, Baron Roland
von Eötvös. He claimed that he had determined the equality of $m$ and $M$ to one
part in a billion. It took the genius of Einstein to point out that such an equality
could not be an accident. He made the profound assumption that inertial mass
is not merely *equal* to gravitational mass but that they are *identical*. The consequences of this assumption led to the general theory of relativity, and to a new
understanding of the nature of gravitation. Because questions of cosmology and
the origins of the universe depend on theories of gravitation, the question of the
identity of inertial and gravitational mass is still of great current interest. In 1966,

figure 5-10    The two different aspects of mass. (Left) it is
a measure of the reluctance of an object to change velocity and (right) it measures the ability of an object to attract another object gravitationally.

Robert H. Dicke, at Princeton, demonstrated the equality to one part in one hundred billion.

From now on, we will assume the identity of inertial and gravitational mass and will use the symbol, $m$, for either property. Gravitational potential energy thus becomes $mgh$, and kinetic energy becomes $\frac{1}{2}mv^2$.

2   In Question 4-4, we raised another troublesome point about gravitational potential energy. How can that energy be proportional to the height $h$? This would mean that there would be no escape from the earth or any other gravitational body. Now that we have defined force, we can also see that if gravitational potential energy is proportional to the height, the force remains constant, independent of the height. This would mean that no matter how far away you got from the earth, your weight, or the attraction of the earth, would remain the same. Since rockets *can* escape from the earth, and since we know that the weight of objects depends on their distance from the earth, we know that something must be wrong with our original definition of gravitational potential energy. Our experiments leading to that definition all took place close to the surface of the earth and applied only to small changes of height. For large distances, the potential energy curve must gradually fall off something like the curve shown in Fig. 5-11. Such a curve would be plausible because it predicts that it takes only a finite amount of energy to escape from the earth.

Let us raise again the choice of a zero point for gravitational potential energy. Although the choice is arbitrary, it is convenient for many purposes to choose the zero point at a distance so far from the earth that the gravitational influence of the earth is negligible. In Fig. 5-11 the slope of the curve at large distances is almost zero. Therefore, the gravitational force at large distances is almost zero, which is what we would expect, and which agrees with observations. If we say that the gravitational potential energy from the earth is zero at infinity, or at least at a distance many times the diameter of the earth, then the curve in Fig. 5-11 must be shifted downward as shown in Fig. 5-12. Note two points about that graph. First of all, the horizontal axis is chosen to be the distance from the *center* of the earth. What goes on at smaller radii inside the earth, we will consider in the next chapter. Secondly, choosing the zero point at a large distance means that the potential energy at any smaller distance is negative. As we saw in Chapter 4, the absolute value of the potential energy is meaningless. All that is important experimentally are the *changes* in potential energy. Even so, it makes sense to talk about our potential energy here on the surface of the earth as being negative. Such wording implies that we are trapped, or bound here, by the gravitational attraction. Indeed we are. In effect, we are in a hole, or a gravitational potential

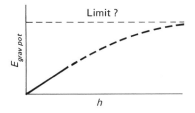

figure 5-11   Proposed dependence of gravitational energy on height.

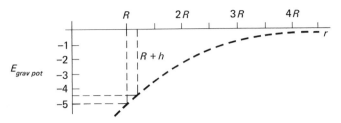

figure 5-12   In this plot of gravitational potential energy as a function of distance from the center of the earth, $r$, the zero point is chosen to be where $r$ is infinite. Therefore, at other points the energy is negative. "$R$" signifies the radius of the earth.

energy well. It takes energy to get out, and that is just what happens when the rockets are given enough kinetic energy; they escape.

The actual shape of the curve in Fig. 5-12 has been known since the work of Newton in the late 17th century. The curve is such that: $E \propto -(1/r)$ where $r$ is the distance to the center of the earth.

*Question 5.10*    From the graph, draw up a table of values of $E$ and $r$ and $1/r$, and show that the curve has been drawn so that $E \propto 1/r$.

Because of one objection to the implications of our original graph of gravitational potential energy, we seem to have arrived at another, entirely different, kind of curve. We started out with $E_{grav} \propto h$, and now are proposing that $E_{grav} \propto 1/r = K/r$ (where $K$ is the proportionality constant). How can these two entirely different expressions be related? Once again we appeal to the fact that an object's potential energy is not of importance, by itself. The only meaningful quantity is the change of potential energy during a particular displacement. The question is, what does each of our two different descriptions predict about the change in potential energy as an object goes from some point, such as the surface of the earth at radius $R$, to a slightly higher point, such as $(R + h)$. These two points are noted on the graph in Fig. 5-12. Note that it is plausible that the curve is almost a straight line for small increases in the distance from the earth's center, $r$.

Algebraically the change in gravitational potential energy is given by:

$$\Delta E_{grav.\ pot.} = E_{final} - E_{original} = \left(\frac{-K}{R + h}\right) - \left(\frac{-K}{R}\right)$$

Multiplying the first term by the quantity, $R/R$, (equal to one); and the second term by the quantity, $(R + h)/(R + h)$, (also equal to one) yields the common denominator, $(R + h)R$, in the expression

$$= \frac{-KR}{(R + h)R} - \frac{-K(R + h)}{(R + h)R}$$

Putting the equation in this form allows the following simplifications to be made:

$$-K\left(\frac{R - (R + h)}{(R + h)R}\right) = K\frac{h}{R^2 + Rh}$$

For values of $h$ which are small compared to $R$ the following approximation can be used:

$$K\frac{h}{R^2 + Rh} \longrightarrow \frac{K}{R^2}h$$

In the approximation, the product, $Rh$, in the denominator, has been ignored compared with $R^2$. To justify this approximation, consider that the radius of the earth is 4,000 miles. Even if we are concerned with a fall of one mile, the product $Rh$ is only 4,000 whereas $R^2 = 16,000,000$. For displacements small compared with the radius of the earth, it appears that the change in potential energy is approximately proportional to the displacement. The proportionality constant, $K$, necessarily has something to do with the earth itself. But that is exactly the situation with our first expression for gravitational potential energy. That earlier expression, $E = mgh$, is apparently an approximation good for small displacements.

Newton not only proposed the $1/r$ relationship for gravitational potential energy, but also specified the nature of the constant, $K$. Since we know that the gravitational potential energy of an object is proportional to its mass, we have $E \propto -(m/r)$. Newton proposed that even as the earth attracts an object such as an apple to it, so *all* objects must attract each other. If the potential energy is proportional to the mass, $m$, of the object, it seems plausible that it must also be proportional to the mass, $M$, of the earth, or whatever the other body is. Therefore: $E_{grav} \propto mM/r$. The proportionality constant must have something to do with the nature of gravitation itself. Its symbol is $G$. The final expression then is:

$$E_{grav\ pot} = -G\frac{mM}{r}.$$

In this expression, mass appears as the "charge" or source of the gravitational influence.

The *change* in potential energy as an object moves upwards a short distance, *h*, from the surface of the earth can now be written:

$$\Delta E = G\frac{mM}{R^2}h.$$

The approximation is

$$\Delta E_{grav\ pot} = mgh.$$

*Question 5.11*  How is the proportionality constant, *g*, related to the mass and radius of the earth and the universal gravitational constant, *G*?

    The force exerted by a system as it tends to reduce its potential energy has been defined to be equal to the negative slope of the potential energy curve. For the approximation $E = mgh$, the force of gravity, or the weight of an object, is equal to *mg*. Since, according to this approximation, the potential energy curve is a straight line, the slope is constant and the force, or weight, must be the same everywhere. Such a requirement doesn't agree with the facts. Let's see how the gravitational force depends on the distance from the center of the earth if the potential energy is proportional to $-(1/r)$. We have already seen the solution to this problem in our treatment of the energy of two cylindrical magnets near each other. If potential energy is proportional to $1/x$, then the force between the two parts of the system is proportional to $1/x^2$. The force of gravitational attraction between two objects must therefore be equal to $G(mM/r^2)$. The force falls off rapidly as the distance between the objects increases.

    This equation is true only for spheres that are separated from each other by a distance greater than the sum of their radii. In other words, it does not apply to an object inside a sphere such as an object inside the earth. Most of the objects to which the formula is applied are spheres, since usually we are concerned with the gravitational interactions of planets or stars.

    The formula is approximately true for any two objects, regardless of their geometry, provided that separation distance between their centers, *r*, is much larger than their individual sizes.

*Question 5.12*  The moon's distance from the earth is about 240,000 miles. The radius of the earth is about 4,000 miles. What is the weight of 1 kilogram, due to the influence of the earth, at a distance equal to the orbital radius of the moon?

    The constant, *G*, is called the *universal gravitational constant*. It is quite different from *g* in that it does not depend on any properties of the earth.

It is called "universal" because, so far as we now know, it applies to any two objects anywhere in the universe. That was the point of the story that Newton told about watching an apple fall. Most people at that time still thought that celestial objects obeyed one set of rules, and earthbound objects obeyed another. Newton had realized that the same kind of interaction that existed between the apple and the earth must exist between the moon and the earth or between any two other objects. The value of the constant can be determined by substituting known values for the force on an object, the mass of the earth, and the radius of the earth. For instance, if $m$ is chosen to be 1 kilogram, then the force acting on it, or its weight, at the surface of the earth is equal to 9.8 newtons. The catch is, although the radius of the earth is fairly easy to measure, how does one measure the mass of the earth?

*Question 5.13*   Make a rough estimate of the value of $G$ by assuming that the earth is a sphere with a constant density equal to that of surface rock. The radius of the earth is $6.4 \times 10^6$ meters. The volume of a sphere is $\frac{4}{3}\pi r^3$. The density of surface rocks is about equal to 2,500 kilograms per cubic meter.

The unknown factor in our calculation is the value for the density of the earth. We could only guess that the average density is equal to the density of rocks on the surface. As a matter of fact, the average density is over twice that. To determine this fact, note that if we know the value of $G$, we can then find the mass of the earth from the gravitational formula. The experimental determination of $G$ is equivalent to weighing the earth! That is just what the British chemist and physicist Henry Cavendish did in 1798. Instead of measuring the attraction between an object and the huge mass of the earth, he measured the extremely small force between two objects in the laboratory. His experimental arrangement is shown in Fig. 5-13. He could control and measure the masses of the attracting objects and the distance between them. He could also measure the very small force required to make the suspension twist through the same angle produced by the gravitational attraction between the spheres. He then solved for the unknown value of $G$. Knowing $G$ and the force of 9.8 newtons exerted by the earth on one kilogram at the surface of the earth, he could calculate the mass of the earth itself. The value of $G$ is very small:

$$G = 6.67 \times 10^{-11} \text{ newton} \cdot \text{meter}^2/\text{kilogram}^2$$

*Question 5.14*   Assuming that the gravitational equation for spheres approximately applies to human bodies if they are far enough apart, calculate the physical attraction between a 70 kilogram boy and a 50 kilogram girl if they are 10 meters apart.

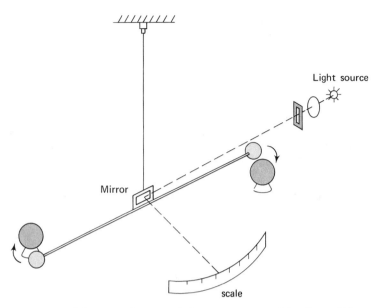

figure 5-13  Schematic diagram of the apparatus used by Cavendish "to weigh the earth". Two spheres are suspended at the ends of a torsion bar, which is hung by a thin fiber. The slight attraction between the suspended spheres and the larger stationary spheres produces a torque that rotates the bar. The effect is amplified by shining a light on a mirror mounted on the axis of rotation. A small angular displacement can thus produce a large linear displacement of the light beam on a scale.

## 5.6   another kind of interaction—electrostatic

We have explored experimentally three kinds of interactions between objects caused by springs, magnetic fields of a particular shape of magnet, and gravitation. In Chapter 1 we claimed that there are only four basic interactions—*gravitational, electromagnetic, strong nuclear,* and one called simply, *weak.* Analysis of the last two of these requires more complicated theory than we can use here. Analysis of the electromagnetic interaction is also complicated because the effects depend on the relative motion of the electrically charged objects.

Depending on these motions the effect can be purely magnetic, purely electrostatic, or a combination of these. For instance, visible light is produced by the electromagnetic interaction and so are all the ordinary chemical forces. The atoms are held together by means of electromagnetism; and consequently, the spring interaction that we studied, as well as the cylindrical magnets, is really a special case of the electromagnetic phenomenon.

We will only analyze here one aspect of electromagnetism, known as *the electrostatic interaction,* for future use in our studies concerning the structure of atoms in Chapter 8. The word *static* means stationary. This particular interaction occurs between electric charges that are stationary with respect to each other. Furthermore, we will describe the rules for only the simplest geometry—where each of the charges is considered as existing at a point, rather than being spread out over a surface.

In such a case, the equation describing the electrostatic interaction follows similar rules to the gravitational. However, instead of the source of the influence being the mass of the object, *m*, the source is an electric charge, *q*. One additional complication is that electric charges can be positive or negative: $+q$ and $-q$. (The atomic electron carries a negative charge; the atomic nucleus carries a positive charge.) The electrostatic potential energy between two charges is given by the relation:

$$E_{elec.\ pot.} = K \frac{q_1 q_2}{r}$$

where *r* is the distance between the charges, and *K* is a proportionality constant. Compare this equation with the one for gravitational potential energy derived on page 119.

$$E_{grav.\ pot.} = -G \frac{mM}{r}$$

Both expressions depend in the same way on the separation distance, *r*. Both depend on the product of two "charge" strengths. The "charge" of the gravitational field is mass; the "charge" of the electrostatic field is electric charge. If these electric charges are measured in terms of the standard unit, called the *coulomb*, the value of the proportionality constant is:

$$K = 8.988 \times 10^9 \frac{\text{newton} \cdot \text{meter}^2}{\text{coulomb}^2}$$

$$K \approx 9 \times 10^9 \ \text{N} \cdot \text{m}^2/\text{C}^2$$

Notice that the formula for *electrostatic* potential energy does not have a negative sign. If the two charges are both positive or both negative, they will *repel* each other. With gravitation, there is only one type of "charge" (positive mass). Two masses *attract* each other. Their potential energy is negative because they trap each other.

*Question 5.15*  Can there be a similar situation with electric charges?

The electrostatic force between two electric (point) charges can be calculated from the potential energy expression. If the potential energy depends on $1/r$, the force depends on $1/r^2$. Therefore, the electrostatic force is given by:

$$F_{electrostatic} = +K\frac{q_1 q_2}{r^2}$$

This equation is known as Coulomb's law, and is named after its discoverer, Charles Coulomb, a French physicist (1736–1806).

*Question 5.16*    What would be the electrostatic force between two positive charges, each with a strength of one coulomb, if they were held ten meters apart? Is your answer physically reasonable?

The electric charges on the basic subatomic particles are very small. The electron, for instance, carries a charge of only $1.6 \times 10^{-19}$ coulomb. Nevertheless, these small charges hold atoms together, and in turn, hold you and us together.

## summary

We have seen that graphs, of potential energy as a function of distance, contain a great deal of information about the interactions involved. Although the potential energy of a system *at one point* gives no indication about the subsequent motion of the system, the graph of potential energy indicates whether or not the object will start to move, and in what direction. This tendency is characterized by the slope of the potential energy curve. The force exerted by a system in reducing its potential energy is defined to be equal to the negative slope of the potential energy curve, $F = -(\Delta E/\Delta y)$. One joule per meter is called a newton of force.

For the original graph of gravitational potential energy, $E_{grav. pot.} = mgh$. Since, according to this, a graph of $E$ versus $h$ is a straight line, the slope of the curve is constant and the gravitational force equals $-mg$. This force, the weight of an object, is directed downward. The potential energy of a compressed spring is proportional to the square of the compression: $E_{spring. pot.} = \frac{1}{2}(kx^2)$. In this case, the force exerted by the spring is proportional to the amount of compression and in the opposite direction, tending to restore the spring to its relaxed condition, $F = -kx$. In the case of the repulsion between two cylindrical magnets, the potential energy is approximately proportional to one over the distance between the pole pieces: $E_{mag. pot.} \propto 1/x$. The force of repulsion is proportional to $1/x^2$.

Work is given the technical definition of force times the distance through which the force is exerted: $F \cdot \Delta x$. By definition, this equals the change in potential energy of the system upon which the force is exerted. The definition was extended to apply to kinetic energy as well as potential energy, since a force exerted on an object might increase its velocity as well as its potential energy of position. An alternative definition of force then appears. The change in potential energy of a system, divided by the distance during which it took place, is equal to the change of momentum of the system, divided by the time needed for the change. Therefore, $F = -(\Delta(mv)/\Delta t)$.

Two unusual features of gravitational potential energy were noted. First, an object's inertial mass, $M$, and its gravitational mass, $m$, were shown to be proportional to each other. Experiments have shown that $m = M$ to a very high precision. We assume that the two properties are identical and use $m$ as the symbol for mass. Secondly, gravitational potential energy cannot be exactly proportional to the distance above the earth or else rockets would not be able to escape from the earth. The more accurate expression is that the gravitational potential energy is proportional to one over the distance, $r$, from the center of the earth: $E_{grav.\ pot.} = -G(mM/r)$. This expression gives the same answer as does the original one for the change of potential energy of an object moving within a short vertical distance near the surface of the earth. Gravitational force is not constant, but falls off rapidly with distance from the center of the earth: $F = -G(mM/r^2)$. The universal gravitational constant, $G$, has a very small numerical value, but can be measured in the laboratory. Once the value of this constant is known, the mass of the earth can be calculated directly from the formula for gravitational force and the known weight of a kilogram mass.

The electromagnetic interaction is responsible for most of the forces that we experience in everyday life. These are the forces exerted by springs, air pressure, muscles, magnets, and electricity, among others. One special case of the electromagnetic interaction is relatively easy to analyze. When two point electric charges, $q_1$ and $q_2$, do not move with respect to each other, their potential is given by:

$$E_{electrostatic\ potential} = K\frac{q_1 q_2}{r}$$

where $r$ is the separation distance. If $r$ is measured in meters, $q$ in coulombs, and $E$ in joules, then $K \approx 9 \times 10^9$ newton $\cdot$ meter$^2$/coulomb$^2$. The force in newtons between the two charges is:

$$F_{electrostatic} = +K\frac{q_1 q_2}{r^2}$$

These equations are similar in form to the gravitational energy and force equations. However, *electrical charge* can be positive or negative; *mass* is only positive and produces only forces of attraction. Therefore, the potential energy of two *masses* is negative because they attract each other and are trapped.

The force between them has a negative sign, implying that they will be drawn together, resulting in smaller and smaller separation distances, *r*. In the *electrostatic case*, if both $q_1$ and $q_2$ are of the same sign (both positive or both negative), they repel each other. The force between them has a positive sign, implying that they will be shoved apart to greater separation distances. Their potential energy is also positive, since they must have been given energy to force them together.

*Answers to Questions*

5.1 (a) If your mass is only 50 kilograms you weigh 110 lbs., or approximately 500 newtons.

(b) When $E_{grav.}$ is plotted *vs* height, *h*, the curve is a straight line rising to the right with positive constant slope. The force, therefore, is negative, which means that it is in the direction of smaller *h*. Since that means "down", gravitational force is in the direction that it should be.

5.2 The slope of the energy curve is small for small distortions, *x*, and is large for large values of *x*. This is qualitatively correct—restoring force is small for small distortions, and large for large distortions of the spring. Since the slope is positive the restoring force of the spring must be negative. This means that as the spring is given larger distortions to the right, the restoring force is in the opposite direction, to the left.

5.3 The force curve that you plot should be approximately a straight line, starting from the origin, with a negative slope.

5.4 The object being pulled upward by the spring is also being pulled downward by gravitation. If the spring potential energy is reduced, the gravitational potential energy would be increased. The object will stay at a point such that if it moves either up or down, the sum of gravitational potential energy and spring potential energy will increase.

5.5 (a) For small separations the slope is very steep, and for large separations the slope is very small. Therefore the force between the repelling magnets must be large when they are close together, and small when they are further apart. Since the slope is negative, the force must be positive which means that it is in the direction of larger separation.

(b) The curve of force versus separation distance for the magnets should have very large values for small *x* and fall off rapidly as *x* increases.

5.6 If the magnets attract each other, they will be in a trapped (or bound) condition when close together. Usually, this situation is represented by assigning negative potential energy to the system. The graph of potential energy for the magnets would then look like Fig. 5-6, except that the curve would be flipped over so that all of the values would be negative. Since the slope would then be positive, the force would be negative, implying that it was in the direction of smaller displacements. In other words, the magnets would be attracted to each other.

**5.7**   If an upward force exactly equal to its weight is exerted on an object, then the object will either remain at rest, or in motion up or down with constant velocity. If a slight extra pull upward is exerted on an object (even just for a moment), it will start moving up. That upward velocity can be *maintained* by exerting an upward force just equal to the weight.

**5.8**   The one-half is an averaging factor. The average force necessary to distort a spring a distance *x*, is only one-half of the force exerted by the spring at the distance *x*. The work done in getting the spring to that condition is equal to the average force times the distance through which the spring was distorted.

**5.9**   If we call the upward direction positive and the downward direction negative, then the *final* momentum, $(Mom)_f$, of the ball is equal to $+1$ kg · m/sec. The original momentum, $(Mom)_o$ was equal to $-1$ kg · m/sec. The force is equal to:

$$F = -\frac{(Mom)_f - (Mom)_o}{\Delta t}$$

$$= -\frac{+1 - (-1)}{0.1}$$

$$= -20 \text{ newtons.}$$

Since the force is negative, it is directed downward, which is the direction we would expect.

**5.10**   One way to demonstrate that $E \propto 1/r$ is to form a fourth column in your table of values which should contain the quantity $E$ divided by $1/r$. If $E \propto 1/r$, then $E/(1/r) = Er = $ constant. If $E$ times $r$ equals a constant, all the values in your fourth column should be the same.

**5.11**   Equate the two expressions for gravitational potential energy: $G(mM/R^2)h = mgh$. Then, divide both sides of the equation by $m$ and $h$: $g = G(M/R^2)$. The constant, $g$, is thus clearly related to the mass and radius of the earth.

**5.12**   At a distance equal to the orbital radius of the moon, an object is about 60 times further from the center of the earth than it is when it is on the surface of the earth. Its weight due to the earth is therefore less by a factor of $60^2 = 3,600$. A kilogram mass out there would therefore weigh about $1/360^{th}$ newton.

**5.13**   The volume of the earth is equal to $\frac{4}{3}\pi r^3 = \frac{4}{3}\pi(6.4 \times 10^6 \text{ m})^3 \approx 1 \times 10^{21} \text{ m}^3$. Therefore, $M \approx 2.5 \times 10^{24}$ kg. Since on the surface of the earth one kilogram weighs approximately 10 newtons:

$$F = G\frac{mM}{R^2} = 10\,N = G\frac{(1)(2.5 \times 10^{24})}{(6.4 \times 10^6)^2}$$

$$G = (10^1)(4 \times 10^{13})/(2.5 \times 10^{24}) = 1.6 \times 10^{-10}.$$

This value is over twice as large as the experimental value. This means that the earth must have a larger mass than we assumed, which means that its average density is greater than the density of surface rocks.

5.14 $\quad F = G\dfrac{m_1 m_2}{R^2} = 6.7 \times 10^{-11} \left[\dfrac{(70)(50)}{10^2}\right] = 2.3 \times 10^{-9}$ newtons.

The boy and girl are scarcely attracted to each other.

5.15 Yes, two electric charges can trap each other, if one is positive and the other negative. Then their product is negative, which is the correct sign for a potential energy well.

5.16 According to Coulomb's law the force would be:

$$F = K\dfrac{q_1 q_2}{r^2}$$

$$= \left(9 \times 10^9 \ \dfrac{\text{newton} \cdot \text{meter}^2}{\text{coulomb}^2}\right)\left(\dfrac{1 \ \text{coulomb} \times 1 \ \text{coulomb}}{(10 \ \text{meter})^2}\right)$$

$$= (9 \times 10^9 \ \text{newtons})\left(\dfrac{1}{100}\right) = 9 \times 10^7 \ \text{newtons}$$

The answer is not reasonable because it represents a tremendous force. Remember that one kilogram weighs ten newtons. Therefore this force is equal to the weight of $9 \times 10^6$ kilograms. One thousand kilograms weighs about as much as our ton. Therefore, this force is about 9,000 tons. It would be almost impossible to mount electrodes to withstand such a force. Besides, before the two isolated coulombs of charge of opposite sign could be accumulated, the potential energy between the electrodes would be so great that the charge would discharge as a lightning bolt.

handling the phenomena

1. *A Check of Hooke's spring law*
According to the definition of force, developed in this chapter, the restoring force exerted by a distorted spring is equal to $-kx$. It is a simple matter to determine whether or not any particular spring satisfies this requirement. (Some do not, in which case their stored energy is not equal to $\frac{1}{2}kx^2$.)

Hang one end of a spring from a firm support and place a meter stick beside the spring. Then add weights, one at a time, until the spring is stretched to about one and a half times its original length. The amount of stretch, $x$, should be expressed in meters. The force, $Mg$, will be in units of newtons if $M$ is expressed in kilograms and the constant, $g$, has the value of 9.8 newtons/kg. Measure at least five points; draw up the data in columnar form, and then graph $F$ on the vertical axis versus $x$ on the horizontal axis. If $F \propto x$, the curve will be a straight line. The slope of the curve will be equal to $k$ in units of newtons/meter.

Note, incidentally, that the restoring force is the one exerted by the spring—not the force exerted by the hanging weights. In equilibrium, however, these are equal in magnitude, although opposite in direction. Since the restoring force is always in the opposite direction from the spring distortion, Hooke's law should have a negative sign on it.

$$F_{restoring} = -kx$$

If the distortion, $x$, is positive (to the right), the restoring force will be negative (to the left).

### 2. *Determining the force law between cylindrical magnets*

You can measure the force exerted by two opposing cylindrical magnets as a function of their separation distance. One magnet should be permanently fastened so that it remains stationary; the other should be mounted in such a way that it is constrained to move only along a straight line away from the permanent magnet. One way to do this is to mount the moving magnet on an air track glider. Another way is to mount the cylindrical magnets in a smooth groove. In any case, the moving magnet must be kept from twisting sideways and yet should be free to move without much friction along the straight line. With the air track glider, for instance, care must be taken that the glider does not twist up or down into the track.

Use a spring force meter to measure the force of repulsion for at least five different separation distances. Make a table of $F$ vs $x$. In a third column list the values for $1/x$, and in a fourth column list values for $1/x^2$. On a graph plot $F$ on the vertical axis, versus the values for $1/x^2$ on the horizontal axis. Since only proportionalities are involved, units are unimportant. For instance, suppose that you measure $x$ in centimeters. For a separation distance $x$ of 2 cm, $1/x = 0.5$, and $1/x^2 = 0.25$. For convenience, you could scale this up by 100 and call the value 25. Choose scales for the axes so that the points are spread out on the graph. If $F \propto 1/x^2$, the points should fall on a straight line through the origin. Since this particular force law is only approximately true for real cylindrical magnets, you should not expect the data to fall perfectly in line.

### problems

1  What is the weight of this book in newtons? If you raise it one meter from floor to table, how much work do you do? What is the resultant increase of gravitational potential energy?

2  Suppose you have a spring that you could use to test your arm strength. When you pull on it as hard as you can, you stretch it 5 cm. What is a reasonable value of the spring constant, $k$?

3  A size D flashlight dry cell stores about 14,000 joules. A coiled car spring can be compressed 10 cm by a force of 5000 newtons (one quarter of a car's weight). If this spring can be compressed as much as 20 cm, how does its energy storage ability compare with that of the dry cell?

4  Draw a sketch of the potential energy graph of a car in a valley. On this sketch,

show graphically the relative strength and *direction* of the gravitational force on the car when it is at the bottom of the valley and at a point on the hillside on either side.

5  Suppose you are riding in a car that changes its speed smoothly from 30 miles per hour to 0 in 0.3 second. Your body is fastened by seat belts, but it has to stop as rapidly as the car does. In metric units, 30 miles per hour equals about 15 meters per second. Will your seat belts hold? Suppose that you and they can sustain a force equal to five times your weight (5 G's).

6  Explain, in your own words, how we obtained two different formulas for gravitational potential energy, and also explain how the two are related.

7  The mass of the moon is $7.3 \times 10^{22}$ kilograms and the distance between earth and moon is $3.8 \times 10^8$ meters. The mass of the sun is $2.0 \times 10^{30}$ kilograms and the distance between earth and sun is $1.5 \times 10^{11}$ meters. What is the ratio of gravitational attraction between sun and earth, and the moon and earth?

8  Suppose that in another universe gravitational mass is equal to the square of the inertial mass. Compare the rate of fall of two objects dropped from rest from the same height at the same time. Assume that the inertial mass of object $A$ is one, and the inertial mass of object $B$ is two. What is the gravitational mass of each? What is the weight of each? After a fall of one meter, how much gravitational potential energy of each is turned into kinetic energy? Then what is the velocity of each at that point? How does this compare with the situation in our universe?

# some applications of a few powerful concepts

Knowing how the potential energy of a system depends on its shape or distortion gives us a great deal of information about what the system might do in the future. In this chapter we want to apply these principles to everything from planets to atoms. We will start out by taking another look at what we have already found out about gravitational potential energy. This information is powerful enough to provide many answers to questions about rockets, weight, and tides. Then we will take a look at rotational kinetic energy. Since this type of energy depends on position within the rotating system, there is a force associated with it. When we see the nature of that force we will have the key to learning many other things about the sun, the earth, and satellites. Finally, we will take a look at other kinds of potential energy situations and shall use these to make a model of the microstructure of matter.

## 6.1  of gravity and human bondage

We live in a potential energy well. We are bound to the earth by gravitational attraction and cannot escape from the earth without turning a large amount of chemical energy into gravitational potential energy. The diagram of Fig. 6-1 portrays the human condition. Zero potential energy is chosen to be at a point very far from the earth. At any closer distance, the potential energy

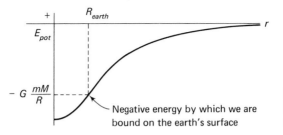

figure 6-1    Gravitational binding energy inside and outside the earth.

is negative. For distances larger than the radius of the earth, the value for the potential energy satisfies the equation

$$E_{grav.\ pot.} = -G\frac{mM}{r} \tag{6-1}$$

As we have seen, this leads to a gravitational attraction between an object with mass, $m$, and the earth with mass, $M$, equal to:

$$F = -G\frac{mM}{r^2} \tag{6-2}$$

These expressions are derived under the assumption that the two objects attracting each other are spheres and are further apart than the sum of their two radii; in other words, they are not penetrating each other. If the objects are not spheres, but are very far apart compared to their size, the formulas are approximately correct. For spheres, $r$ is the distance between centers, and so in the case of the earth and an object near it, $r$ is the distance between the object and the center of the earth.

If a hole could be drilled down to the center of the earth, we would find that the potential energy and force of attraction inside the earth follow rules different from those that we have just described. It is obvious that they could not possibly be described by the formulas above because these formulas require that the attractive force on an object becomes infinite when $r$ goes to zero. It would also take an infinite amount of energy to extract an object up the hole from the center of the earth. The actual situation is that once an object is inside the earth the net attraction on it is due only to that part of the earth at a smaller radius than the object. The attraction caused by all the material in the shell at a radius larger than the position of the object cancels out. See Fig. 6-2. The result of this is that the force on an object inside a sphere *with uniform density* is proportional to the distance to the center. (Actually, as has been demonstrated earlier, the earth must have a greater density at the center than it does at the crust.)

$$F \propto -r \tag{6-3}$$

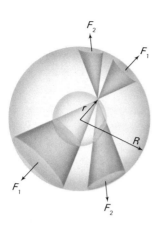

figure 6-2   Explanation of force acting on an object inside the earth. The outward force on object at $r$ due to section of earth close to it, is exactly balanced by force in opposite direction caused by section outside that radius on the opposite side of the earth. Since the gravitational pull of each region in the vicinity of the object is exactly balanced by the pull of a region on the opposite side, only the gravitational pull of the material inside radius $r$ is effective.

This expression is the same as the force rule for an ordinary spring. The potential energy is therefore proportional to $r^2$. Notice how reasonable the expression is. It says that the force on an object at the center of the earth where $r$ equals zero would be zero. That makes sense. The earth pulls equally in all directions and all the forces cancel. The potential energy curve for the region inside the earth is shown in Fig. 6-1. The potential energy does not go to infinity, but of course remains finite even at the center of the earth.

*Question 6.1*   If the earth were not rotating, if it had uniform density, if a hole could be bored through it along a diameter, and if there were no air resistance, what would be the motion of a ball dropped into the hole? Remember that the force on the ball follows the same rule as the force exerted by a spring.

The potential energy graph illustrates the way that we are trapped on the surface of the earth. The potential energy of an object with mass, $m$, on the surface of the earth (where the distance to the center is $R$) is equal to:

$$E_{trapped} = -G\frac{mM}{R} \qquad (6\text{-}4)$$

To escape the earth's pull, we must provide enough positive kinetic energy so that the total energy is equal to zero. If an object is thrown straight upward, it will rise until all of its kinetic energy is changed to potential energy. If at that point the potential energy is zero, then the object will not fall back again. The condition for escape is:

$$E_{kin} + E_{pot} = 0$$

$$\tfrac{1}{2}mv^2 - G\frac{mM}{R} = 0 \tag{6-5}$$

We can now solve this equation for the velocity necessary to escape. Note first of all that we can divide through by the mass of the object so that it cancels out. The escape velocity is independent of the mass of the object. It is the same velocity for an atom as it is for a rocket. To solve for the escape velocity, $v_e$, we transform the equation as follows:

$$\tfrac{1}{2}mv_e^2 = G\frac{mM}{R}$$

Multiplying both sides of the equation by two, and then taking the square root gives:

$$v_e = \sqrt{G\frac{2M}{R}}. \tag{6-6}$$

*Question 6.2*    To two significant figures, the values for the constants are: $G = 6.7 \times 10^{-11}$ newton meter$^2$/kilogram$^2$); $M_{earth} = 6.0 \times 10^{24}$ kilograms; $R_{earth} = 6.4 \times 10^6$ meters. Find the escape velocity in meters per second. One mile/hour equals 0.45 m/sec. What is the escape velocity in miles/hour? Does this agree with newspaper accounts of the velocities achieved by rockets?

Suppose that we give a rocket enough energy so that it can escape from its potential well and be free of the earth. Is it then free to go wandering around the solar system, perhaps to journey to some other star? Remember that we are also under the sun's gravitational influence. The graph in Fig. 6-3 shows what a terrible hole we are in. Even when we escape from the

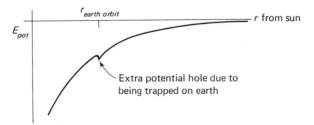

figure 6-3    A comparison of the gravitational potential energy wells produced by the earth and the sun.

earth's gravitational pull, we have hardly begun our long journey away from the sun.

*Question 6.3*   Compare the depth of the sun's gravitational potential well at the distance of the earth's orbit, with the depth of the gravitational potential well on the surface of the earth due to the earth's gravitation. In the first case, for instance, the potential energy is given by our standard formula where $r$ is the distance between earth and sun, equal to $1.5 \times 10^{11}$ m. The controlling mass in this case is that of the sun, which is equal to $2.0 \times 10^{30}$ kg. Before substituting numbers, form the ratio by placing one expression over the other. Some of the factors will cancel out.

As you know, the giant Saturn rockets that took the astronauts to the moon were much larger than the rockets in which the astronauts landed on the moon and used to return to earth. The binding energy of an object at the surface of the moon, and the escape velocity from it, are considerably less than the earth values. We can compare these binding energies by taking their ratio:

$$\frac{E_{pot\ earth}}{E_{pot\ moon}} = \frac{-G\frac{(mM_{earth})}{R_{earth}}}{-G\frac{(mM_{moon})}{(r_{moon})}}$$

$$= \frac{M_{earth}\,r_{moon}}{M_{moon}\,R_{earth}} = \frac{(6 \times 10^{24})(0.27\,R_{earth})}{(7 \times 10^{22})(R_{earth})} \approx 20$$

*Question 6.4*   What is the ratio of the escape velocity from the earth to the escape velocity from the moon?

## 6.2   rotational energy

One way or another, we all go in circles a fair share of the time. There is energy associated with this motion and it can be expressed in several different ways. First of all, if we are traveling with a velocity, $v$, we have a kinetic energy equal to $\frac{1}{2}mv^2$. As shown in Fig. 6-4, if we are traveling in a circle at constant velocity, our velocity is equal to the circumference, $2\pi r$, divided by the period, $T$, which is the time that it takes to make one full revolution. Substituting this expression for velocity into the expression for kinetic energy:

$$E_{kin.\ rot.} = \tfrac{1}{2}mv^2$$

$$= \tfrac{1}{2}m\left(\frac{2\pi r}{T}\right)^2 = \tfrac{1}{2}m\frac{4\pi^2 r^2}{T^2} \tag{6-7}$$

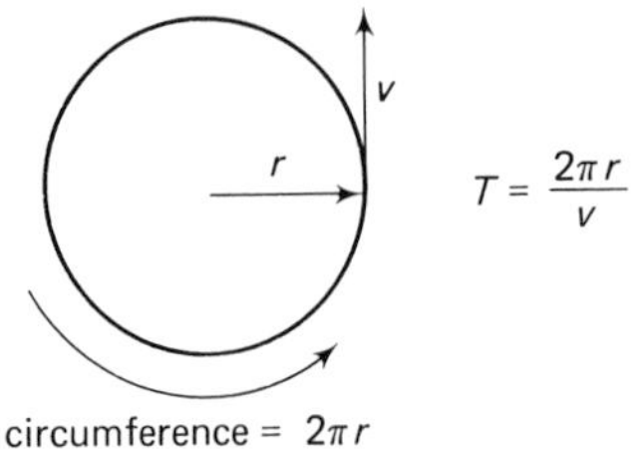

$$T = \frac{2\pi r}{v}$$

circumference $= 2\pi r$

figure 6-4   Relationships among the velocity, $v$, radius, $r$, and the period of rotation, $T$, for an object in circular rotational motion.

By this change of variables, we have expressed rotational kinetic energy in terms of position. As an example, suppose that you are on a merry-go-round. No matter where you are riding, the period of rotation, $T$, is the same. However, the kinetic energy that you possess, because of the rotation, is proportional to the square of your distance from the axis: $E_{rot.} \propto r^2$ or, $E_{rot.} = \frac{1}{2}Kr^2$. Notice that all reference to motion has disappeared. *The rotational kinetic energy has assumed the form of a potential energy dependent on the position, r.* With any such potential energy, it is necessary to determine the zero point and whether the energy is positive or negative. In this case the determination of the zero point is simple: when $r = 0$, $E_{rot.} = 0$. The question of whether $E_{rot.}$ is positive or negative for $r > 0$ depends on who has to do work to restore the object to zero position. If the system automatically tries to restore the object to zero position, then $E_{rot.}$ is positive. (This is like the situation with a stretched spring.) If an outside agency has to provide energy to move the object back to zero, then $E_{rot.}$ must be negative. Experience shows that this latter is the case. On a merry-go-round it requires human effort to work your way from the rim to the center. Therefore $E_{rot.}$ is negative, just like gravitational binding energy.

$$E_{rot.} = -\tfrac{1}{2}m\,\frac{4\pi^2}{T^2}\,r^2$$

Except for the negative sign, this rotational kinetic energy has the same mathematical form as the potential energy of a spring: $E_{spring} = \frac{1}{2}(k)x^2$. There is a restoring force associated with the potential energy of a spring: $F = -kx$. We could follow the same algebraic derivation and find the force associated with the rotational energy. Since

$$E = \tfrac{1}{2}(k)x^2 \longrightarrow F = -(k)x, \text{ then}$$

$$E_{rot.} = -\frac{1}{2}\frac{(m4\pi^2)}{(T^2)}r^2 \longrightarrow F_{rot.} = +\frac{(4\pi^2 m)}{(T^2)}r. \tag{6-8}$$

According to this analysis, since a change of position, $r$, leads to a change in energy, $E$, then there must be a force associated with the rotating system. It is proportional to the radius, $r$, and is in the direction of the radius away from the center. This is called the *centrifugal* (or center fleeing) *force*. It is the force experienced by an object in a rotating system.

Two different words are used to describe this radial force of rotation. We have talked about a *centrifugal* force acting away from the center of a circular path. Under other circumstances, it is useful to talk about a *centripetal* force with the same magnitude but acting inward along the radius. How can there be both a force acting in and an equal force acting out on an object going in a circle?

There is not really any paradox here; it is all a question of the reference frame from which the force is being observed. As shown in Fig. 6-5, if you are looking down on a car that is going around a curve, then you realize that there must be a force *in* towards the center, a centripetal force, in order for the car to stay on the road. The driver, on the other hand, who is in the reference frame of the rotating object feels a centrifugal force pulling him away from the center. There is no controversy here *so long as observations from the two different reference frames are not mixed.* It makes no sense, for instance, to think that the centrifugal force is the reaction to the centripetal force.

*Question 6.5*    Describe the force acting on the driver in a drag race, first from the reference frame of the driver and secondly, from the reference frame of an observer outside the car. Is the force on the driver in the forward direction or the backward direction? (In a drag race, the driver starts a powerful car from rest and accelerates as fast as possible.)

If you swing a stone that has been attached to one end of a string around in a circle, the source of the centripetal force is evident. Your hand exerts a tension on the string which pulls the stone in toward the center. If the string

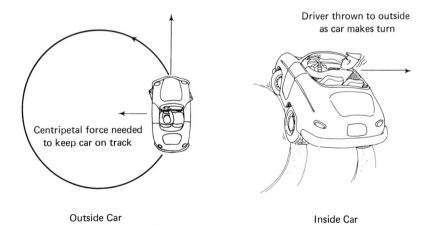

Centripetal force needed to keep car on track

Driver thrown to outside as car makes turn

Outside Car                          Inside Car

figure 6-5    Whether the radial force in circular motion is centripetal or centrifugal depends on your reference frame.

breaks, or if you let go, the centripetal force disappears and the stone flies off at a tangent. The source of the centripetal force controlling the circular motion of planets around the sun is not so obvious. It was Newton who demonstrated that gravitational attraction provides the necessary centripetal force. By equating the two we can obtain an important relationship between the period and radius of revolution of a planet or satellite. Gravitational force provides centripetal force.

$$-G\frac{mM}{r^2} = -\frac{4\pi^2 m}{T^2}r$$

$$-G\frac{mM}{r^2} = -\frac{4\pi^2 m}{T^2}r$$

$$-G\frac{M}{r^2} = -\frac{4\pi^2}{T^2}r$$

$$-\frac{GM}{r^3} = -\frac{4\pi^2}{T^2}$$

$$-4\pi^2 r^3 = -GMT^2$$

$$\diagup r^3 = \diagup \frac{GMT^2}{4\pi^2}$$

$$r^3 = \left(\frac{GM}{4\pi^2}\right)T^2 \tag{6-9}$$

By performing these algebraic transformations we have derived one of Kepler's laws of planetary motion. Kepler, who lived at the time of Galileo in the late 16th and early 17th century, had analyzed the data of an astronomer named Tycho Brahe. His measurements of planetary positions, all taken without optical telescopes, were sufficiently precise for Kepler to use in testing various models of the solar system. The main controversy in astronomy at the time was between the Copernican theory, that the planets go around the sun, and the Ptolemaic theory, that all the heavenly objects rotate around the earth. Kepler's analysis showed that the orbits of the planets around the sun can be described as ellipses. The elliptical paths for the planets are very close to circles. Kepler's analysis also demonstrated the law that we have just derived—that $r^3 \propto T^2$ for all of the planets around our sun, a relationship that is true for ellipses as well as circles.

Question 6.6    The average orbital radius, $r_{orbital}$, and the period, $T$, of the planetary year are given for five planets in the table below. These values are all given in terms of the corresponding ones for the earth, $r_e$ and $T_e$ respectively. Demonstrate that Kepler's relationship holds for these planets.

| Planet | $r_{orbital}$ | $T$ |
|---|---|---|
| Mercury | $0.39\ r_e$ | $0.24\ T_e$ |
| Venus | $0.72\ r_e$ | $0.61\ T_e$ |
| Earth | $1.00\ r_e$ | $1.00\ T_e$ |
| Mars | $1.53\ r_e$ | $1.88\ T_e$ |
| Jupiter | $5.22\ r_e$ | $11.85\ T_e$ |

(The easiest way to demonstrate Kepler's relationship is to modify the table, and prepare a fourth column with values of $r^3$, a fifth column with values of $T^2$, and a sixth column with values of $r^3/T^2$. If $r^3 \propto T^2$ then $r^3/T^2$ is a constant.)

Kepler's relationship, between the orbital radius and period, holds for the satellites of any gravitational body. For instance, this relationship must be satisfied by all the satellites of the earth. In this case, the mass of the earth must be used in the proportionality constant. Notice that the mass of the satellite cancels out of the relationship. An artificial earth satellite, for instance, has a particular period at a given radius regardless of its mass. By substituting values for $G$ and $M$ into the formula we could find the period for a circular orbit of a particular radius. With even less arithmetic we can compare the orbits of two satellites—for instance, an artificial one and the moon. In this case the proportionality constant cancels out and we have:

$$\frac{r^3_{moon}}{r^3_{sat.}} = \frac{T^2_{moon}}{T^2_{sat.}} \tag{6-10}$$

To find the relationships between two orbits, we can use any units that we want so long as we use the same units for the moon as we do for the satellite. For instance, for the period of the moon we can use 27 days, and for its distance from the earth, 240,000 miles.

*Question 6.7*   What is the period, in hours, of an earth satellite that is only one or two hundred miles above the surface of the earth? Its orbital radius, in this case, is approximately $\frac{1}{60}$th that of the moon's.

A satellite with particular usefulness for worldwide communication systems is one that hovers over the same point above the earth. To do this, the plane of its orbit must be the same as that of the equator, and the period of its revolution must be 24 hours. To find how far away such a satellite must be, we can use the Keplerian relationship to compare its orbit with that of the moon.

$$\frac{r^3_{sat.}}{r^3_{moon}} = \frac{T^2_{sat.}}{T^2_{moon}} \qquad \frac{r^3_{sat.}}{240,000^3} = \frac{1^2}{27^2}$$

Notice that here we are expressing both periods in days and both radii in miles. As seen from the earth, the moon's period to get back to the same place *with respect to the stars* is approximately 27 days. (Due to the motion of the earth-moon system about the sun, to get to the same relative position with respect to the earth and sun takes a day longer. It is this latter time from full moon to full moon that we normally think of as the period of the moon. For our purposes in comparing satellites, the shorter period, which is one complete revolution around the earth, is the one to use.)

$$\frac{r_{sat.}}{240,000} = \frac{1^{\frac{2}{3}}}{27^{\frac{2}{3}}} = \frac{1}{9} \qquad r_{sat.} = 27,000 \text{ miles}$$

Such a hovering satellite is approximately 23,000 miles above the surface of the earth at the equator. Several of these are now in orbit relaying messages from one earth location to another.

A satellite can be thought of as having two different kinds of energy. It has kinetic energy due to its motion, and it also has gravitational potential energy because of its position near the controlling body. Its total energy is the sum of these two. For circular orbits:

$$E_{total} = (\tfrac{1}{2}mv^2) + \left(-G\frac{mM}{r}\right)$$

For a satellite in circular orbit there is a special relationship between its velocity, $v$, and its orbital radius, $r$. This relationship results from the condition that the centripetal force is equal to the gravitational attraction.

$$\frac{4\pi^2 m}{T^2}r = G\frac{mM}{r^2}$$

As we saw earlier, for a circular orbit, the velocity and the period are related as follows:

$$T = \frac{2\pi r}{v}$$

By substitution we can now eliminate $T$ from the expression for the equality of the gravitational and centripetal force:

$$G\left(\frac{mM}{r^2}\right) = \frac{4\pi^2 m}{\left(\frac{4\pi^2 r^2}{v^2}\right)}r$$

$$G\frac{mM}{r^3} = 4\pi^2 m \frac{v^2}{4\pi^2 r^2}$$

$$G\frac{mM}{r^3} = m\frac{v^2}{r^2}$$

$$G\frac{M}{r} = v^2$$

We now have an expression for $v^2$ of a satellite in orbit:

$$v^2 = \frac{GM}{r}$$

The kinetic energy of a satellite can now be rewritten in terms of the radius of its orbit:

$$E_{kin} = \tfrac{1}{2}mv^2 = \tfrac{1}{2}m\frac{GM}{r} \tag{6-11}$$

The total energy is equal to:

$$E_{total} = \left(\tfrac{1}{2}G\frac{mM}{r}\right) + \left(-G\frac{mM}{r}\right) = -\tfrac{1}{2}G\frac{mM}{r} \tag{6-12}$$

For an object in circular orbit, the kinetic energy is equal in magnitude to one-half the potential energy. Since the kinetic energy is positive and the potential energy is negative, the total energy is still negative.

*Question 6.8*   What is the significance of a negative total energy?

If an object in motion loses energy, it usually slows down. Peculiar things happen in orbit, however. Suppose that a satellite in an orbit close to the earth gradually loses energy due to collisions with air molecules. We know very well that the satellite will fall lower and lower and eventually spiral into the earth. The radius of its orbit will get smaller. Look what happens to kinetic energy, however. If the radius, $r$, gets smaller, the kinetic energy increases. That means that the satellite will have a larger velocity. This is indeed the case. As the satellite loses energy, the orbital radius shrinks, and the satellite's velocity gets larger and larger.

*Question 6.9*   What is the relationship between the velocity of the moon and the velocity of a satellite close to the surface of the earth?

Within the last few years we have heard a great deal about the unusual dynamics of life on an artificial satellite. The most troublesome feature is the effect of weightlessness on the astronauts. Whether or not they are truly

Three trajectories of rockets fired from a high tower (above the earth's atmosphere). The first two are suborbital.

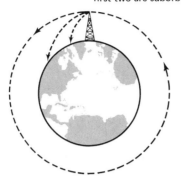

figure 6-6    A satellite is in "free fall" toward the earth, but the surface of the earth curves away so that the satellite does not hit the earth.

weightless is largely a matter of semantics. From the reference frame of an earth observer the satellite and its occupants are attracted by gravity. That influence, of course, is what provides the centripetal force necessary to keep the satellite from flying off at a tangent. If the satellite is only 100 miles or so above the earth, the gravitational force is not much smaller than it would be on the surface. In the reference frame of the *satellite*, the astronauts experience a *centrifugal* force which exactly balances the gravitational attraction. According to any experimental method that they can use—such as observing a suspended plumb bob, for example—they are weightless. The words that we use to describe this situation are relatively unimportant. One way to describe what is happening is to say that the satellite and its occupants are falling freely at all times toward the earth. Under these circumstances, the astronauts are weightless with respect to the satellite and everything in it. But if the satellite is forever falling toward the earth, why doesn't it hit it? As shown in Fig. 6-6, the satellite would strike the earth if it did not have enough velocity, or if the earth were flat. In one sense, the satellite is falling toward the earth, but the surface of the earth is curving away from under the satellite.

*Question 6.10*    Here is another problem of satellite life. Suppose that you are in the same orbit as another satellite, but slightly behind it. What happens if you fire your rocket engines behind you in order to catch up with the first satellite?

## 6.3   angular momentum

Almost all of our study of the physical universe in this text has grown out of our discovery of some conserved properties. It has been possible to define various types of energy so that under certain simple circumstances energy is always conserved. We saw that the momentum of an isolated system always remains the same. There is another type of momentum, which is associated with rotational motion. The law of conservation of angular, or rotational, momentum is as important as any of the others. The conservation of angular momentum dominates the dynamics of galaxies, planets, and subnuclear particles. Besides, the quantity has some very special properties whose significance is only partially understood at this time.

The angular momentum of an object is always given with regard to its distance from some axis of rotation. In Fig. 6-7, for example, an object, with mass, *m*, is going past the point labeled *O*, and might or might not swing around it. It has an ordinary momentum equal to *mv*, while its angular momentum *with respect to the axis* is *mvr*. (The radius of rotation used to define angular momentum is always taken perpendicular to the velocity.) It might not be obvious that an object has any rotational tendencies merely because it is passing by some axis. Suppose one skater glides by another skater,

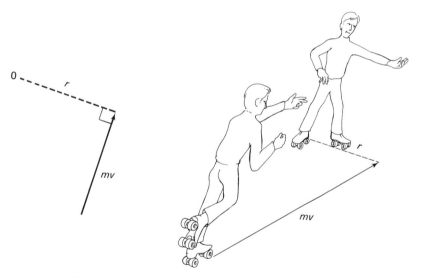

figure 6-7   The diagram shows the definition of angular momentum about an axis. Even though the skater moves in a straight line and no circular motion is apparent, still he has angular momentum about the point, *O*. If he were to clasp hands with a skater stationed at point *O*, both of them would whirl around and the angular momentum would be evident.

approaching to a closest (perpendicular) distance, $r$. There appears to be no rotation involved. If, however, the two skaters clasp hands as the first one goes by, then the two are clearly going to start whirling around each other.

Figure 6-8 is a diagram of an elliptical orbit around the sun. This could be the orbit of a comet, but it also represents an exaggeration of the elliptical orbit of any of the planets. The angular momentum at three particular points is noted. We claim that the angular momentum of such an isolated system must remain constant. Yet clearly the distance between satellite and controlling body continually changes.

*Question 6.11*    How can $mvr$ for a satellite remain constant if $r_1 > r_2 > r_3$?

The quantity of angular momentum can be expressed in another way for objects in circular rotation. In this case the velocity is equal to the circumference divided by the period: $v = 2\pi r/T$. Angular momentum can then be expressed in terms of the period of rotation:

$$\text{angular momentum} = mvr \longrightarrow m(2\pi/T)r^2. \tag{6-13}$$

Ordinary linear momentum depends only on an object's mass and velocity. Angular momentum also depends on *where the mass is located in the rotating system*. This definition agrees with everyday observations. Compare the difficulty of stopping two rotating wheels, each of which has the same mass, but which are constructed as shown in Fig. 6-9. One of the wheels has most of its mass concentrated in the rim, like a bicycle wheel. The other has a light weight rim and has its mass concentrated in the axle. If both wheels have the same period of rotation, the angular momentum of the first one is much larger than that of the second. It would be harder to get it going and harder to stop it.

Because linear momentum is conserved, a completely isolated object cannot change its velocity. Although angular momentum is also conserved, an isolated object *can* change its rotational velocity. There is no paradox here. Every ice-skater knows how to do it. You can demonstrate the effect yourself by sitting on a rotating stool with your hands stretched out. Spinning slowly like that, part of your mass is at a large radius. Suddenly draw your hands in, reducing the radius of that part of your mass. Angular momentum, which is

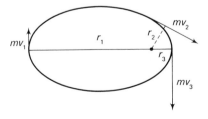

figure 6-8    If the angular momentum $mvr$, is to remain constant, the velocity $v$, must change as the comet orbits the sun.

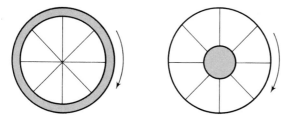

figure 6-9 Two wheels with the same mass. The one on the left has its mass concentrated in the rim, in the manner of a bicycle wheel. The one on the right has its mass concentrated near the axle. The one on the left is much harder to start or stop than the other.

equal to $m(2\pi/T)r^2$, must be conserved, and yet you have just reduced the radius, $r$, of some of your mass, $m$. To compensate for this, your period of revolution, $T$, must also be reduced. You will start spinning faster. If: $m(2\pi/T_1)r_1^2 = m(2\pi/T_2)r_2^2$ and: $r_2 < r_1$ then $T_2 < T_1$. For more details about doing this experiment and the one described in the question below see Section 2 of Handling the Phenomena at the end of the chapter.

*Question 6.12* Suppose that you are swinging a stone on a string as shown in Fig. 6-10.

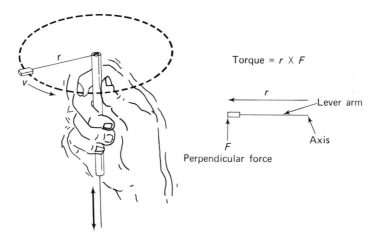

Torque = $r$ X $F$

figure 6-10 This is a device for whirling an eraser in a circular orbit with variable radius. Pulling down on the string reduces the radius without any torque being applied to the system. (Since the force in this device is applied to the eraser along the radius, no change in angular momentum can be produced. You cannot make something rotate by pulling along the radius arm. A "torque", which produces a change in angular momentum, is the product of a lever arm radius and a force perpendicular to the lever arm at that radius.)

By drawing the string through the tube, you can reduce the radius of rotation to one half of its original value. What happens to the period of rotation?

Not only is the magnitude of angular momentum conserved in an isolated system, but the direction also remains constant. The direction of ordinary momentum is obvious enough. But it is not so obvious that angular momentum also has a direction. Of course, a wheel can be described as rotating clockwise or counterclockwise as viewed from a particular direction, but its angular momentum can also be defined as being along the direction of the axis of rotation. See Fig. 6-11. This is an arbitrary, but convenient, definition of the direction of angular momentum. The fact that the direction of angular momentum of a rotating wheel is not easily changed is part of the explanation of why a person can balance on a moving bicycle but not on a stationary one. If a person tries to tip a spinning bicycle wheel, not only is there a resistance to the tipping motion, but the effect is at right angles to the direction that you would normally expect it to be in. Bicycle riders know this intuitively. If they are riding "no hands" and want to turn to the right, they have to tip slightly off balance to the right. For suggestions about studying these gyroscopic effects see Section 3 of Handling the Phenomena at the end of the chapter.

A spinning top is pulled by gravity in such a manner that it would topple over if it were motionless and in the same position. Instead, it stays roughly at the same angle to the horizontal, but precesses around the vertical axis in such a way that the axis of rotation traces out a conical shape. Our earth does the same thing as it spins. Its axis of daily rotation is tipped so that the equatorial plane of rotation is not the same as the orbital plane of annual revolution. The sun exerts a force on the slight equatorial bulge of the earth, but this does not eliminate the tilt of the earth. Instead, the earth's axis precesses with a period of twenty-six thousand years. The north star, to which the axis is now pointing, is not the same as the pole star in other eras, as

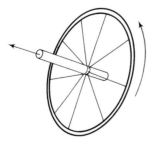

figure 6-11    The arbitrary (but conventional and useful) direction of angular momentum is directed along the axis of rotation in the following way. If you were to curl the fingers of your right hand in the direction of rotation, your thumb would point along the axis in the conventional direction.

figure 6-12   The tilt of the earth's axis, with respect to its plane of revolution around the sun, precesses with a period of 26,000 years.

illustrated by Fig. 6-12. (The reason that "it is the dawning of the age of Aquarius" is also due to precession. The apparent position of the sun at the time of the vernal equinox, with respect to the constellations of the zodiac, is continuously moving. Since there are twelve constellations, or "houses" in the zodiac (including Aquarius) it takes this point an average of 2000 years to move through one house. Actually we have another few hundred years before the constellation Aquarius is reached, and the "age of Aquarius" actually begins.)

There is one other feature about angular momentum that sets it apart from any of the other quantities we have studied so far. It has a natural minimum unit. There is no such natural unit for mass, length, time, or force. So far as we know, we can divide lengths up into smaller and smaller amounts indefinitely. There is no known basic unit of mass from which all other masses must be derived. A rotating system, however, cannot have just any amount of angular momentum. There is a basic unit. In terms of our standard units of kilograms, meters, and seconds, it has the fantastically small value of approximately $1 \times 10^{-34}$ kilogram meter$^2$ per second. A spinning object can change its angular momentum by one of those units, or two of those units, or three of those units, but not by a fraction of that unit. In this way angular momentum is very similar to electric charge, which also must exist in integral multiples of a basic unit. Of course, the basic unit of angular momentum is so very small that we do not see its influence in the rotation of human sized objects. You can change the speed of a bicycle wheel as smoothly as you wish. For atoms or subatomic particles, however, the effect is vital. The magnitude or direction of their spin changes only in jumps.

*Question 6.13*   If the unit of angular momentum were $10^{34}$ times larger, what effect would you notice while riding a bicycle?

## 6.4   other forms of potential energy

Our basic technique of describing potential energy in terms of graphs gives us a very potent tool. So far we have described the simple cases of potential

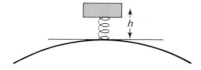

figure 6-13   A weight compressing a spring is subject to two potential energies—that of gravitation from the earth, and that of the spring.

energy due to magnets, springs, rotation, and gravitation. If an object is subject to several influences, we may be able to describe its behavior with a potential energy graph even though the algebraic description might be very complicated. As a simple example, consider a weight supported by a spring as shown in Fig. 6-13. The whole system consists of the object, the earth, and the spring. There is gravitational potential energy due to the separation of the object from the earth, and there is potential energy of the spring caused by the compression of the spring. The potential energy curves due to these two effects are shown in Fig. 6-14. The horizontal axis shows height above the floor, and the vertical axis records the potential energy. The graph for the gravitational potential energy under these circumstances is a straight line, given by the equation $E_{grav.\ pot.} = mgh$. The potential energy due to the spring is described graphically here, in a slightly different way than before. Previously, we measured the extension of the spring from the undistorted position. In this case, the zero on the horizontal axis would mean that the spring is pushed into the floor. For small values of extension, $h$, the spring potential energy would be very large since the spring would be highly compressed. At some particular value of $h$ the spring is in its normal (undistorted) condition, and so the potential energy must be zero. For larger values of $h$, the spring is

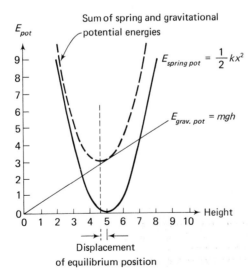

figure 6-14   Combined potential energy curves for an object on a spring in a gravitational field. The weight of the object has displaced the equilibrium position of the spring from its original location.

stretched and its potential energy again rises. The curve is a displaced parabola. The potential energy of the whole system must be the sum of these two separate potential energies. That curve is shown by a dotted line on the graph. It is obtained simply by adding the energy coordinates of both curves at each particular height. The object on the spring is subject to both influences and the resultant force on it is equal to the negative slope of the combined energy curve.

*Question 6.14*  At what height will the resultant force on the object be zero? What is the direction of the force on either side of that point?

A real object supported on a spring in this manner will vibrate up and down if it is displaced from its equilibrium position. Its total energy situation is then very much like that of a swinging pendulum. At the extreme points of its motion, up or down, its velocity is zero and its energy is completely due to its position in the gravitational and spring system. As it passes through the equilibrium position, its velocity is maximum and its energy can be considered exclusively kinetic. (If additional energy is given to the system, the amplitude of oscillation will increase. The object will go up higher, down further, and will go through the equilibrium position faster than before.) Graphically, the division of energy between kinetic and potential can be represented as shown in Fig. 6-15. The height of the horizontal line is proportional to the total energy of the system. As the object moves from its lowest

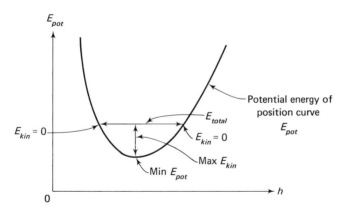

figure 6-15  Potential energy well for an object oscillating on a spring. At a given level of total energy (kinetic plus potential), the object can slosh back and forth in the well, transforming energy from kinetic to potential and back again.

to its highest position, its total energy remains the same. However, the potential energy of position decreases as the object rises, until it is at a minimum at the equilibrium position of the spring. As the object continues to rise, the potential energy of position starts to increase, reaching a maximum at the spring's longest extension. The difference between the horizontal total energy line and the potential energy curve represents the kinetic energy, which is at a maximum at the equilibrium point, and at a minimum where the total energy line and potential energy curve intersect.

Any object found in a potential energy well must behave something like this. Without knowing anything about the forces that bind atoms together to form solid material, we can propose that the potential energy curve of a bound atom must look something like that shown in Fig. 6-16. The potential energy of the atom in the system, $E_{pot.}$, is plotted along the vertical axis, and the distance, $r$, from a neighboring atom is plotted along the horizontal axis. If the two atoms are part of a solid there must be some form of attraction between them. We cannot know at this point whether this attraction depends on the inverse first power, inverse second power, or some higher inverse power of the separation distance. But the general form of the potential energy curve must look something like that in the diagram. For large separation distances, the potential energy must be zero. For smaller separation distances the binding energy must get greater and greater. Since it is energy of trapping, we assign negative values to it. If the two atoms get too close together there must be some sort of force that keeps them from merging into each other. We have represented this region on the graph with a sharply rising potential energy. Between the regions of repulsion and attraction there must be an equilibrium position of separation between the two atoms.

The restoring forces acting on the atom are also shown in Fig. 6-16.

*Question 6.15*    Justify the shape of the force diagram for the atom in a potential well. How is it related in each region to the potential energy curve?

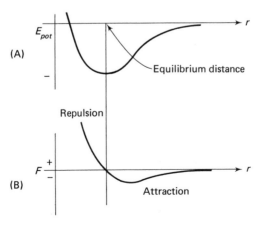

figure 6-16    Plausible graph of potential energy for a bound atom (A). The force experienced by the atom as it moves within the potential well (B), can be derived by graphing the slopes at all the points on the potential energy graph.

Our plausible description of atomic binding energy can be carried even further. What will an atom do if it is displaced slightly from its equilibrium position? In analogy with what happened to the block on the spring, we can claim that the atom will vibrate around its equilibrium position. Its total energy will then be part kinetic and part potential, but still negative, implying that it is bound. Suppose that even more energy is fed into the atomic system so that its total energy becomes zero, or positive. According to our picture, the atom would slosh out of its potential well and would no longer be trapped.

## summary

In this chapter we have applied and expanded our definition of potential energy and momentum. We saw that our description of gravitational potential energy could explain many facts about rockets and satellites. The graph of gravitational potential energy, drawn so that the energy is negative, illustrates the way that we are trapped in a potential well. To escape from the surface of the earth we must give the rocket a kinetic energy equal to the negative potential energy that the rocket originally possessed. Its total energy therefore becomes zero, which means that it can escape the earth.

By equating kinetic energy to gravitational binding energy, the escape velocity was calculated. We also applied the formulas for gravitational potential energy and gravitational force to compare the escape velocities of objects on the earth and the moon.

Our standard expressions for energy also apply to rotational motion. It is useful to express these in terms of the period of rotation. Rotational kinetic energy then appears in the form of energy as a function of position. A force is associated with this energy. This is the centrifugal force, which is proportional to the radius and directed along the radius away from the center. Whether the force is centripetal or centrifugal depends upon the frame of reference in which the description is made.

We can find a fundamental law describing satellite motion by equating gravitational force to the centripetal force necessary to make the satellite go in a circular orbit. This law, first derived by Kepler (though not in this way), applies to the planets around the sun or the satellites around our earth. We derived other rules for planetary motion by specifying the kinetic and potential energy of an object in orbit. The kinetic energy of a satellite is positive and equal in magnitude to one half of its gravitational potential energy. The total energy for the bound satellite is therefore still negative. The definite relationships between kinetic and potential energy of a satellite lead to unfamiliar dynamics.

Another conserved quantity of an isolated system is angular momentum. This quantity, which is defined as the linear momentum, $mv$, times the perpendicular distance, $r$, of the orbit to some axis of rotation, has both magnitude and directional properties. Both are conserved in an isolated system. The response of a rotating system to outside forces gives rise to gyroscopic effects.

The graphical method of describing the potential energy of a system can be usefully applied even to situations where we do not know the exact forces involved. For instance, we can predict the general behavior of an atom bound in a solid by making very general assumptions about the way it is trapped.

*Answers to Questions*

**6.1** Since the force on the ball is proportional to its displacement from the center of the earth, and in a direction pointed toward the center of the earth, the ball will behave just as if it were a weight bobbing back and forth on a spring. Starting from the surface of the earth, the ball will travel faster and faster toward the center. The force, meanwhile, will be getting smaller until at the center of the earth the force is zero. At that point, the velocity and kinetic energy are at their maximum values, but the potential energy is at its minimum value. As the ball passes by the center, the restoring force starts pulling it back, eventually slowing it down to a stop by the time it reaches the surface on the opposite side. Its velocity and kinetic energy are once more zero at that point, but its potential energy is now at its maximum value. The ball will fall back down, repeating the excursion over and over again.

**6.2**

$$v_e = \sqrt{(6.7 \times 10^{-11}) \left(\frac{2 \times 6.0 \times 10^{24}}{6.4 \times 10^6}\right)} = \sqrt{1.3 \times 10^8} =$$

$$1.1 \times 10^4 \text{ meters/sec} = 25{,}000 \text{ mph.}$$

**6.3**

$$\frac{E_{sun}}{E_{earth}} = \frac{G\left(\frac{mM_s}{R_s}\right)}{G\left(\frac{mM_e}{R_e}\right)} = \frac{M_s R_e}{M_e R_s} = \frac{(2.0 \times 10^{30})(6.4 \times 10^6)}{(6.0 \times 10^{24})(1.5 \times 10^{11})} = 14$$

**6.4** Since the binding energy is equal to the escape energy, which is equal to $\frac{1}{2}mv^2$, the ratio of escape velocities must be equal to the square root of the ratio of binding energies.

$$\frac{v_{earth}}{v_{moon}} = \sqrt{\frac{E_{pot.\ earth}}{E_{pot.\ moon}}} = \sqrt{20} \approx 4.5$$

**6.5** When a drag race driver starts up, he is thrown back against his seat and feels that there is a force on him in the *backward* direction. An observer outside the car would say that the back of the car's seat had exerted a strong force on the driver to propel him *forward*. Both observers would agree about the magnitude of the force and each observer is correct about its direction so long as he specifies the reference frame from which he is observing the force.

**6.6** Here are the fourth and fifth columns of the table relating orbital radii and periods of the planets. As you can see, the ratio of $r^3/T^2$ is approximately one for all the planets.

| Planet | $r^3$ | $T^2$ |
|--------|-------|-------|
| Mercury | 0.059 | 0.058 |
| Venus | 0.37 | 0.37 |
| Earth | 1.00 | 1.00 |
| Mars | 3.58 | 3.53 |
| Jupiter | 142 | 140 |

6.7

$$T^2_{sat.} = T^2_{moon}\left(\frac{r^3_{sat.}}{r^3_{moon}}\right) = \frac{\overset{9}{27} \times \overset{2}{24} \times 27 \times \overset{2}{24}}{\underset{5}{60} \times \underset{\underset{5}{20}}{60} \times 60} = \frac{27 \times 9}{125} = 1.9 \text{ hrs}^2$$

$$T_{sat.} = \sqrt{1.9} \approx 1.4 \text{ hrs.}$$

6.8   According to our convention, a system with negative total energy is bound in some way. The satellite is, of course, still trapped by its controlling object.

6.9   The velocity of a satellite close to the surface of the earth is much greater than that of the velocity of the moon. Since $v^2 = GM/r$, and the ratio of radii of the moon and a close earth satellite is about 60, then the satellite must have a velocity greater than the moon by a factor of $\sqrt{60} \approx 8$.

6.10   If you increase your kinetic energy in a satellite by firing rocket engines behind you, your total energy becomes greater, which means less negative. The orbital radius therefore becomes larger and your velocity smaller. That is no way to catch up with the rocket ahead of you.

6.11   If the radial distance of a satellite in an elliptical orbit changes, then the velocity must also change. That is just what happens to comets in highly eccentric elliptical orbits. Their velocity close to the sun is very large, but far away from the sun is very small.

6.12   Since the only tension on the string is perpendicular to the velocity of the stone, there is no way to change the stones' angular momentum: the quantity $(2\pi m/T)r^2$ must remain constant in order for angular momentum to be conserved. Therefore, since the expression is divided by four when the radius is halved, the value of $T$ must be divided by a factor of four in order for the whole expression to remain constant.

   Mathematically, the original expression $(2\pi m/T)r^2$ becomes

$$\left(\frac{2\pi m}{T}\right)\left(\frac{r}{2}\right)^2 \quad \text{which equals} \quad \left(\frac{2\pi m}{T}\right)\left(\frac{r^2}{4}\right)$$

In order to make this expression identical with the original, thus conserving momentum, the following operation must be performed:

$$\left(\frac{2\pi m}{\frac{T}{4}}\right)\left(\frac{r^2}{4}\right)$$

This new equation can now be reduced to the original form. Physically, the result is that the frequency of rotation is quadrupled, and angular momentum is conserved.

6.13 The bicycle wheel could rotate only at a particular frequency, or twice that frequency, or three times that frequency, etc. Furthermore, the axis of the wheel could change its angle of orientation only in small steps and could not be lined up at positions in between these angles.

6.14 The object is in equilibrium at the height corresponding to the point where the curve resulting from the summation of the kinetic and potential energy curves has zero slope. On either side of that point the restoring forces are such as to bring the object back to that point.

6.15 At the equilibrium point of the potential energy curve where the slope is zero, the force is zero. To the left of that point, the slope is negative and so the force is positive, indicating repulsion. To the right of the equilibrium point, the slope is positive and so the restoring force is negative, indicating attraction. The force rapidly gets smaller as the atom separates beyond a particular stretch limit.

handling the phenomena

1. *Observing centrifugal force*
As described on page 136, centrifugal forces are present for an observer *in* a rotating reference frame. (An observer outside the rotating system will see these forces as being centripetal.) The magnitude of a centrifugal force on an object with mass $m$ is equal to $m(4\pi^2/T^2)r = m4\pi^2 f^2 r$, where $T$ is the period, $f$ is the frequency, and $r$ is the radius. If the units used are kilograms for mass, seconds for period, number of revolutions per second for frequency, and meters for radius, the force will be in newtons.

You can run a qualitative check on this relationship, and actually see centrifugal forces in action, by installing a simple device on a record player turntable; preferably, one that can be switched to two or more speeds. Take a tall, small diameter (frozen juice type) can, that is open at one end. Scotch tape or glue a rod across a diameter of the closed end so that the rod sticks out at least 20 cm from the center on each side. The rod should be made of lightweight material; it could be a tinker toy dowel or a couple of heavy plastic straws. The length of the rod (about 40 cm) and the height of the can should be such that when the can is placed over the central record spindle the rod can sweep around above any supports or levers adjacent to the turntable. Suspend two plumb bobs from each side of the rod, one at 10 cm from the center, and one at 20 cm. The thread could be a light, flexible string, or a heavy thread, preferably white. The bobs can be made of almost any small dense material; paper clips work well. Make sure that the whole system can rotate freely. A diagram of the apparatus is shown in Fig. 6-17.

A record player can usually turn at 33, 45, or 78 *revolutions per minute*. Figure out approximately what these frequencies equal in revolutions per second. (For these purposes, 33 rpm is approximately $\frac{1}{2}$ rps.) Compare the centrifugal force at a given radius and frequency with the mass of the object. Note that the mass of the object

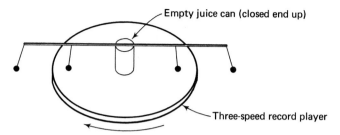

Empty juice can (closed end up)

Three-speed record player

figure 6-17   Simple apparatus for making qualitative observations of centrifugal force. Scotch tape holds the rod to the
top of the juice can.

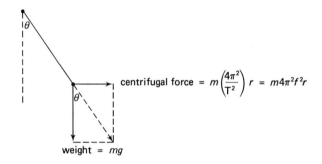

$$\text{centrifugal force} = m\left(\frac{4\pi^2}{T^2}\right)r = m4\pi^2f^2r$$

weight = $mg$

$$\text{tangent } \theta = \frac{\text{opposite}}{\text{adjacent}} = \frac{\text{centrifugal force}}{\text{weight}} = \frac{m4\pi^2f^2r}{mg} = \frac{4\pi^2f^2r}{g}$$

figure 6-18

cancels out:

$$\text{centrifugal force/weight} = m4\pi^2f^2r/mg.$$

In the meter-kilogram-second system, the value of $g$ is 9.8 newtons/kg. (For this
purpose, $g \approx 10$.)

As shown in Fig. 6-18, the angle that the rotating plumb bob makes with the
vertical is given by: tan $\theta$ = centrifugal force/weight. Observe the rotating system
to see if there is qualitative agreement with your calculation. At 33 rpm, is the angle
of the 20 cm plumb bob about twice that of the 10 cm plumb bob? If you double
the frequency, the centrifugal force should increase by a factor of four. If you
increase the frequency by 45/33, approximately what is the increase in centrifugal
force? Do your observations agree with these calculations?

2.   *Conserving angular momentum by speeding up*

As discussed on page 145, the conservation of angular momentum does not require
the conservation of the frequency of revolution. The expression for angular momentum is $mvr$ where $m$ is the mass of the object, $v$ is its linear velocity (along a straight

line), and $r$ is the perpendicular distance to the axis around which the angular momentum exists. If the object is going in a circle, $v = 2\pi r/T = 2\pi rf$, where $r$ is the radius of the circle ($2\pi r$ is the circumference), $T$ is the period of revolution, and $f$ is the frequency, in revolutions per second. Therefore, the angular momentum for an object going in a circle is $m2\pi r^2/T = m2\pi r^2 f$. If the angular momentum remains constant then the radial location of the mass, $r$, can increase if the frequency decreases, or vice versa.

To experience this effect, sit on a stool that can be freely rotated and whirl around slowly with your arms outstretched. Suddenly bring your arms in close to your body, thus reducing the radius of some of your mass. The effect can be greatly increased by holding books or other heavy objects in your outstretched hands. To slow down, providing you haven't fallen off the stool, simply stretch your arms out again. CAUTION: be careful in doing this demonstration. The effect can be so startling that a person can easily fall off or strike some nearby piece of furniture.

The same phenomenon can be seen with the simple apparatus shown in Fig. 6-11. Slip a string through some kind of tube and tie something soft, like an eraser, to one end of the string. Use the tube as a handle to whirl the eraser around in a horizontal circle about one foot in radius. It will be necessary to hold the string coming down the tube with your other hand. (Thus, in your reference frame, you are exerting a centripetal force on the eraser.) Now pull the string so that the radius of the whirling eraser is only half what it was to begin with. What happens to the frequency? Does it double?

Since you pulled on the string, the force that you exerted (having changed direction by 90° upon leaving the tube) was radial, and therefore it could not have changed the angular momentum. Consequently, the angular momentum remained constant and the eraser speeded up. What happened to the rotational kinetic energy? Surely you exerted a force on the eraser in the direction in which it moved—radially inward. Therefore you performed work on the system. Can you account for the extra energy that the system ought to possess?

3.   *Dealing with wheeling*

A sketch of a bicycle wheel is shown in Fig. 6-11, and it is proposed that it is a useful convention to assign the direction of angular momentum to be along the axis of rotation, at right angles to the plane of rotation. Not only is the magnitude of angular momentum conserved in an isolated system, but its direction is conserved, as well. This effect is important in bike riding, among other things. Try to balance on a bicycle when the wheels are not turning!

Gyroscopic effects can be made dramatically tangible by using a bicycle wheel, preferably one equipped with handles on either side of the axle. If no such wheel is available, you can get some idea of the effects by spinning the front wheel, while holding the front end of a bicycle off the ground. Use a bike with heavy wheels (not a racing bike) and let someone else spin it. Note what happens when the wheel is spinning in a vertical plane and you attempt to turn it left or right. Conversely, try to tip the wheel over and watch which way it turns. Note the directions involved and see what differences occur if the wheel is spinning in the opposite direction. Evidently, the direction of a large angular momentum is not easily changed. Furthermore, the applied torque produces a change of direction perpendicular to that expected without prior experience. You can also observe these effects by riding no-hands. Lean to the side and see which way the wheel turns.

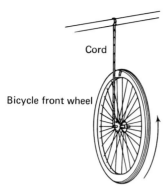

figure 6-19

If you have a wheel liberated from its bicycle frame, suspend it from a stout cord fastened to one end of its axle as shown in Fig. 6-19. Hold the wheel in a vertical plane and spin it vigorously. Now let go and watch the resulting precession. The torque due to its weight, which you might think would cause the wheel to flop down, instead causes it to precess around the axis of suspension. You can increase the torque by adding extra weight to the free handle. What does this do to the rate of precession?

## problems

1   Our atmosphere contains oxygen, nitrogen, and many other gases, but does not contain hydrogen or helium. Since these gases are continually fed into the atmosphere by various processes, they must be able to escape out of the gravitational potential well of the earth. Use the equations for kinetic energy and gravitational potential energy to explain how lighter gases might escape. After all, isn't escape velocity independent of the mass of the object? (At any given temperature, the *average* kinetic energy of any kind of gas molecule is equal to the average kinetic energy of any other kind of gas molecule. In other words, at any temperature where hydrogen and oxygen are mixed: $\frac{1}{2}m_{H_2}v_{H_2}^2 = \frac{1}{2}m_{O_2}v_{O_2}^2$.

2   The mass of the moon is $7.3 \times 10^{22}$ kilograms and its radius is $1.7 \times 10^6$ meters. Find the escape velocity from the moon and compare it with that from the earth.

3   Flywheels are sometimes used to store large amounts of energy for short periods of time (minutes or hours). Suppose that you could construct a flywheel that would fit in an automobile. Could you store enough energy to drive very far? (A gallon of gasoline furnishes about $10^8$ joules, but only about 20 percent of this energy turns into useful propulsion energy.) Assume reasonable values for the radius and mass of the flywheel. (It has to fit in the car and should not have a mass greater than one half that of the car.) For simplicity of applying the formulas, assume that all of the mass of the flywheel is located on the outer rim at one particular radius. The maximum rotational velocity, for safety's sake, should be 3600 revolutions per minute.

4   A person's weight is slightly greater at the north pole than it is at the equator, because the distance to the center of the earth is slightly smaller at the north pole than at

the equator. (The earth has a small equatorial bulge.) There is also another small effect caused by the centrifugal force at the equator. What is the outward centrifugal force experienced by a 60 kilogram person on the equator? Compare this force with the person's weight. (The radius of the earth is $6.4 \times 10^6$ m. Be sure to use seconds for the period.)

5    Suppose that you spin an object in a circle using a spring (or a rubber band) as the radius arm. The system has two forms of energy: that due to rotation, and that due to the stretched spring. Write down the expression for the sum of these two, which is the total energy of the system. Graph the two separate expressions, *making sure that you use the proper sign for rotational energy*. Note that the spring constant, $k$, is related to the value of the rotational constants, because:

$$\text{spring force} = -\text{centrifugal force}$$

$$-kr = -m\frac{4\pi^2}{T^2}r$$

$$k = \frac{m4\pi^2}{T^2}$$

Does your result make sense? (Note that if you tried to walk *in* radially on a merry-go-round, you would have to be working against centrifugal force. If you were fastened to the center by a spring, the restoring force of the spring would always provide the correct force so that you could walk back and forth radially without exerting any other force.)

6    To get a rough idea of how a skater can pirouette, assume a model of a human body as follows: 90 percent of the mass is located on the rim of a cylinder 10 centimeters in radius. The other 10 percent, consisting of arms and hands, can be extended to a radius of 60 centimeters. Suppose that this system is whirling with outstretched arms at one revolution per second, and then the arms are suddenly brought in to the radius of the rest of the mass. What is the resultant rotational velocity? (Note that the total angular momentum of the system is just the sum of the angular momentum of the parts. Also note that since you will form an equation stating that the original angular momentum is equal to the final angular momentum, any units can be used for mass, radius, etc. Those same units will then apply to both sides of the equation. For instance, for the mass of 90 percent of the body, you can use a mass of nine. The mass of the other 10 percent is, of course, one. Call the radius of 10 cm, one; the radius of 60 cm is, therefore, six.)

7    An electron in an outer atomic orbit has a velocity of about $1 \times 10^6$ meter/second. The radius of such an orbit is about $1 \times 10^{-10}$ meter. The mass of an electron is $9 \times 10^{-31}$ kg. Compute the angular momentum of an electron in such an orbit, and compare the value with the basic unit of angular momentum for a rotating system, ($1 \times 10^{-34}$ kilogram meter$^2$/second).

8    On the upper half of a piece of paper, sketch a potential well of the type shown in Figure 6-16. Use a ruler and sharp pencil to find relative slope values of the curve. Plot these values in the bottom part of the graph to show how force on the atom depends on separation distance, $r$. The measuring and plotting technique is the same as that used in answering Question 5-3.

*Chapter Seven*

# heat, temperature, and the conservation of energy

The time of reckoning has come. In the air track experiments described in Chapter 2 we went to great trouble to create simple conditions to study simple collision events. It was possible to define two quantities that were conserved in such collisions—momentum and kinetic energy. However, if the air track gliders had to move against springs, or magnetic fields, or uphill, they lost their kinetic energy. To preserve a conservation law we invented another form of energy, called potential. The various kinds of potential energy were defined so that the sum of kinetic and potential energy would be conserved, but usually only under very artificial conditions. In the real world the air track gliders do slow down, the springs stop bobbing, and even the pendulums eventually come to rest. Where does the energy go? Perhaps it is not conserved. It is certainly not obvious that there are grand conservation laws in the universe. As we shall see, the concept of energy conservation was not proposed in a satisfactory way until the middle of the nineteenth century and possible exceptions have been explored as recently as the 1940's.

What happens to objects when their motion energy and potential energy disappear? Careful measurement shows that they get hotter, or their volume changes, or their state changes from solid to liquid or from liquid to gas. To relate the missing energy to these physical changes, we need a model of the microstructure of matter. If we can show in detail how this microstructure soaks up the gross mechanical energy of the system, then perhaps we can still claim that energy is a conserved quantity. In the process we will find that this model explains other phenomena, such as the production of light, and the nature of chemical interactions. In building this model, our main experimental tool will be not a microscope or an atom smasher, but rather—a thermometer!

# 7.1   heat and temperature

In everyday life we all know what temperature is. It has something to do with how hot or cold it is and we measure that quantity with a thermometer. However, what's the relationship between temperature and heat? What do we mean when we say, "It's not the heat, it's the humidity"? If we "heat up something," do we necessarily raise its temperature?

*Question 7.1*   Write out in a few sentences your own operational definitions of heat and temperature, giving your present understanding of the distinction between them.

Despite the fact that heat and temperature are common household words, it is a fact that most people would have considerable difficulty answering Question 7.1. The following simple experiment illustrated in Fig. 7-1, is useful for demonstrating the distinction between heat and temperature. Suppose that we obtain a small immersion heater like those sold in pet stores for warming tropical fish tanks. First we place the heater in a small glass of water for five minutes. At the end of this time we find that the water feels considerably hotter than before. Now we put this same heater in a bathtub filled with water and again leave it there for five minutes. In this case the tub of water will not seem much hotter than it did before. The degree of "hotness" is what

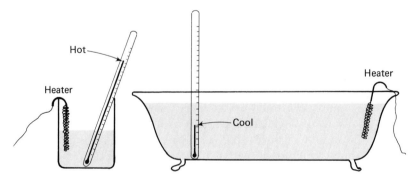

figure 7-1   The same amount of heat that will make the water in a small glass hot, will have much less effect on the "hotness" of the water in a full bathtub.

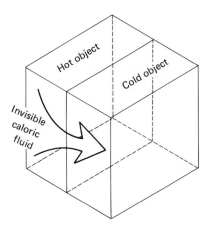

figure 7-2   The caloric theory explained heat transfer as being due to the flow of an invisible fluid from hot to cold objects.

we call temperature, and thus the small glass of water has undergone a much larger temperature change than the bathtub of water. Nevertheless, the amount of energy fed into each system by the electrical heater was the same. Heat and temperature are evidently two different quantities. Heat is energy flowing into (or out of) a system because of a temperature difference. It is often true that adding more heat to an object will raise its temperature, but not always. For instance, if you add heat to boiling water, the temperature of the water remains the same.

It is not obvious that heat is energy moving from one object to another. Many properties of heat suggested to early workers that it consisted of an invisible, and weightless, fluid to which Antoine Lavoisier gave the name "caloric." An elaborate theory of heat was constructed on the basis of a model that pictures the transfer of heat as the flow of caloric. The fact that two objects left in contact soon come to the same temperature was explained by the presumed tendency of the caloric fluid to distribute itself uniformly by flowing from the hotter object (in which it was assumed to be more concentrated) into the colder object. See Fig. 7-2. The observation that substances expand when heated was attributed to the absorption of caloric fluid.

Quantitative experiments were required to refute the fluid theory of heat. The death blow resulted from the inability of the caloric model to explain how heat could be continually produced so long as objects are rubbed together. Was the supply of caloric in an object inexhaustible?

*Question 7.2*   How could the caloric fluid theory have accounted for the production of a small amount of friction heating?

These qualitative ideas about heat and temperature are not very useful unless they can be operationally defined. Since mechanical energy that disappears because of friction often produces a temperature change, we should clearly define how such changes are to be measured.

## 7.2    measurement of temperature

The human body itself is a rather crude instrument for measuring temperature and is useful as a thermometer only over a limited range. The body's unreliability, even in the temperature range where it is most sensitive, can be demonstrated by a familiar trick. Soak your right hand in ice water for a few minutes and at the same time soak your left hand in water as hot as you can stand. Then place both of them in a pan of water at room temperature. Your brain will then receive two conflicting messages. According to your right hand the tepid water is hot, but according to your left hand it is cold.

Devices for measuring temperature are generally called *thermometers*. All of us have had experience in their use and it might seem that no further discussion is necessary. It is fine to know how to read a thermometer but, as with all scientific measuring devices, learning to read the scales or dials is not sufficient. Before employing these tools for scientific purposes one should understand how they function and what their limitations are. A scientist interested in measuring the temperature on the cold side of the moon would find the ordinary mercury thermometer of the earthbound meteorologist as useless as a fever thermometer is to the cook who wants to check the temperature in an oven before popping in a roast.

The thermometers we are all most familiar with consist of a liquid confined in a tube with a bulb at the end. Their operation is based on the fact that the volume occupied by the liquid changes with temperature.

*Question 7.3*    The fractional change in volume for liquid mercury is only about $10^{-4}$ per Fahrenheit degree ($\sim 2 \times 10^{-4}/°C$). The height of the mercury column in a household fever thermometer is observed to change by an amount equal to about one twentieth of the total thermometer length for a one Fahrenheit degree change in temperature. Explain how this magnification of the small change in volume is accomplished.

Other common thermometers are based on the fact that different solids will change in length by different amounts for the same change in temperature. If two equally long strips of different metals are bonded together and heated, the resulting bimetallic strip will bend into an arc, as illustrated in Fig. 7-3, with the metal that expands at a greater rate on the outside. Such a device can be made to change the position of a pointer on a scale which is calibrated to indicate the temperature.

*Question 7.4*    Suppose that the metal that changes in length by the greater amount is always found on the outside of a curved hot bimetallic strip. What happens when such a composite strip is cooled to a temperature below the point at which the two component strips have the same length?

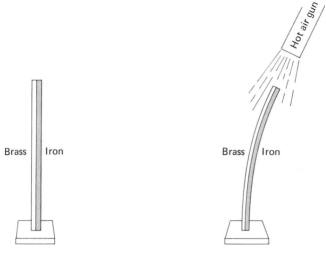

figure 7-3   Effect of temperature change on a bimetallic strip. In this instance the brass expands
more per unit of temperature change than does iron.

Changes in temperature produce other effects. These include variations in
the resistance of a metal to the flow of an electric charge and changes in
the pressure of a gas confined in a fixed volume. Special purpose thermometers
based on these and many other temperature sensitive properties of matter
have been constructed. A key common property of most thermometric devices
is that they respond to temperature in a manner that is both *reversible* and
*reproducible*.

*Question 7.5*   List some effects of changes in temperature that are not reversible.

In order to convert a device that responds to temperature changes in a
reversible, reproducible manner into a thermometer it is necessary to calibrate
it. Suppose we wish to calibrate an unmarked mercury thermometer such as
the one shown in Fig. 7-4a. We must first choose two reference temperatures.
The freezing and boiling points of water are convenient for this purpose.[1]
The bulb of the uncalibrated thermometer is first cooled to the freezing point
of water by placing it in a well stirred ice-water mixture. Sufficient time is
allowed for the mercury to reach temperature equilibrium with the ice-water

---

1. As will be discussed in Chapter 10, the temperature at which a liquid boils (and to a lesser
extent the temperature at which it freezes) is affected by the pressure exerted on the liquid.
It is therefore necessary to specify the atmospheric pressure at the time the calibration is
performed in order to define the two reference temperatures precisely.

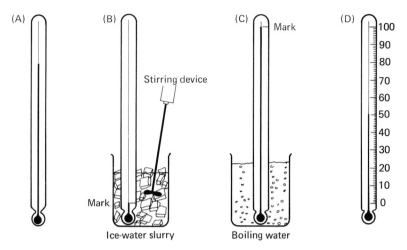

figure 7·4   (A) An uncalibrated mecury thermometer. (B) A reference mark
has been added to indicate the level of the mercury at the freezing point of
water. (C) Second reference mark added to show the mercury level in the
thermometer at the boiling point of water. (D) A fully calibrated
thermometer.

mixture, and the final height of the mercury in the glass tube is marked by
a scratch as shown in Fig. 7-4b. Next, the thermometer bulb is placed in boiling
water and the new equilibrium mercury level is marked with a second scratch,
Fig. 7-4c. To create a temperature scale we arbitrarily label the first scratch
zero and the second 100. The distance between is divided into 100 equal
segments (which we call *degrees*) by marking 99 additional equally spaced
scratches between the two reference points. The final calibrated thermometer
is shown in Fig. 7-4d. Its reversibility and reproducibility can be proven by
demonstrating that the mercury level always returns to the zero and 100 marks
when the thermometer is placed in freezing or boiling water respectively.
We can use this thermometer to record the value of any temperature between
our two reference points by observing the height of the mercury column on
this scale. In fact we can extend the range of our scale below zero or above
100 by continuing to make equally spaced scratches.

*Question 7.6*    The maximum usable range of a mercury thermometer calibrated in the
above manner is about −38 degrees to +356 degrees. Explain.

Suppose we calibrate a second thermometer that is based on the expansion
of a different liquid (alcohol is commonly used), or on a different reversible
temperature effect, in exactly the manner specified above. We again mark

the freezing point of water as zero and the boiling point as 100, and thus the two thermometers are in exact agreement at these two reference temperatures. If we now use these two thermometers to measure some intermediate temperature will they agree? The general answer is *no*. In some cases the difference might be too small to be significant, but in other instances it might be large. When one thermometer reaches the 50° mark it means that the temperature sensitive property on which this thermometer is based is exactly halfway between its value at the boiling and freezing points of water. At this same temperature other thermometer materials might be 45% or 49% or 51% of the way between their zero and 100 degree values, and would therefore indicate a temperature of 45, 49, or 51 degrees. Two thermometers calibrated at the same two reference temperatures will give identical readings between the reference points only if the change in the physical properties on which the thermometers are based are exactly proportional to one another. Does this mean that an official temperature scale must be based on the arbitrary choice of some standard thermometer material? A sophisticated mathematical argument in terms of the "laws" of thermodynamics establishes a temperature scale that is independent of the choice of a temperature sensitive property of a particular substance. We will return to this point at the end of this chapter. This thermodynamic scale is identical to the temperature scale obtained under ideal conditions with a gas volume thermometer. A simple model of such a thermometer, based on changes in the volume of a gas under constant pressure, is illustrated in Fig. 7-5. Ordinary mercury thermometer readings will deviate to a small extent from this physically meaningful temperature scale, but by changing the uniform spacing of the scale markings or by applying an experimentally determined correction formula it is, of course, possible to correct for these deviations.

*Question 7.7*     Suppose you have to calibrate a thermometer that is based on a property $(A)$ which varies in the following way with the temperature $(T)$ as defined by the gas thermometer scale:

| $A$ | 10 | 20 | 40 | 70 | 110 | 160 |
|-----|-----|-----|-----|-----|-----|-----|
| $T$ | 0° | 20° | 40° | 60° | 80° | 100° |

Draw a picture showing the spacing of the markings on the $A$ thermometer scale arranged so that you could read the temperature directly.

Thus far we have used the freezing point of water as the 0° point on our temperature scale and the boiling point of water as 100°. This choice defines the *Celsius* (sometimes still called centigrade) scale with temperatures indicated by a number followed by the symbols, °C, read "degrees Celsius". An outdoor thermometer calibrated on this basis will give negative degree read-

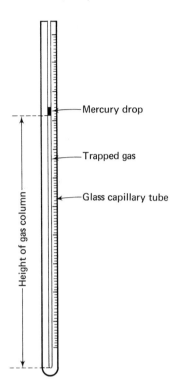

figure 7-5   A simple gas volume thermometer. The height of the gas column as measured by the bottom of the mercury drop will be proportional to changes in the volume of the gas trapped in the glass capillary tube.

ings whenever the temperature drops below the freezing point of water. On the Celsius scale the boiling point of liquified air is about $-192°C$. The gas volume thermometer, although not useful below the temperature at which the gas in the thermometer liquifies, can be used to establish an apparent lowest possible temperature by extending the line drawn through the points on a volume-temperature graph to the point at which the volume of the gas would be zero. See Fig. 7-6. If this is done carefully it is found that this "lowest possible temperature" is $-273.2°C$ independent of the gas employed. This

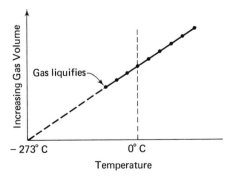

figure 7-6   Graph of the volume $vs$ temperature relationship of a gas, extrapolated to the point where the volume would be zero.

temperature is known as "absolute zero" and has been shown to correspond physically to the condition of matter when no further heat can be extracted from it. By choosing $-273.2$ as the zero point on our temperature scale, and retaining the same temperature *increment* between degrees as is used for the Celsius degree, we obtain the *Kelvin* scale. On this scale the freezing and boiling points of water are $273.2°$K and $373.2°$K, respectively. All other Kelvin temperatures are exactly $273.2°$ higher than their Celsius equivalents.

Most of the countries in the world have officially adopted the Celsius temperature scale and virtually all temperatures in the scientific literature are reported as either °C or °K. In some English-speaking countries (but no longer in England) the Fahrenheit system is still the official one. In this sytem, the freezing and boiling points of water are also used as reference temperatures but they have been assigned the values of $32°$F and $212°$F respectively.

*Question 7.8*   Show that the correct algebraic formula for converting Celsius temperatures to Fahrenheit temperatures is: $T(°F) = \frac{9}{5}T(°C) + 32$. What is the numerical value of absolute zero on the Fahrenheit scale? See Fig. 7-7.

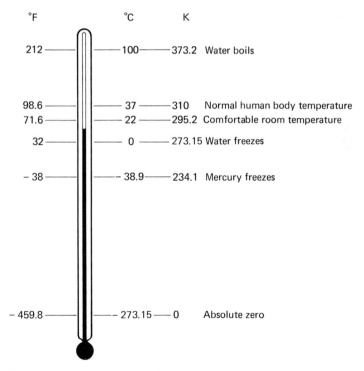

figure 7-7   A comparison of the Fahrenheit (°F), Celsius (°C), and Kelvin (K) temperature scales.

## 7.3   the mechanical equivalent of heat and the conservation of energy

Quantitative investigations of phenomena associated with heat transfer were well underway before the demise of the caloric theory. A unit for specifying a quantity of heat was required for these studies. It was decided to define this unit, which is called the *calorie* (cal), as the amount of heat required to raise the temperature of one gram of water by one Celsius degree. The *Calorie* (note capital *C*) referred to in dietetics is actually a kilocalorie (kcal)—equal to 1,000 calories. The final acceptance of the idea that heat phenomena involve energy changes in the microstructure of matter required a demonstration that a quantitative equivalence exists between the calorie and other units of energy. Experimental and theoretical work directed toward this goal was carried out during the early part of the 19th century by the expatriate American, Count Rumford; by the German physician, Julius Robert Mayer; and, in its most ingenious and elegant form, by the British amateur scientist, James Prescott Joule. A schematic diagram of a device similar to one employed by Joule is illustrated in Fig. 7-8. The mass, $M$, falls a distance, $h$. The gravitational potential energy ($Mgh$) lost by the falling mass becomes the kinetic energy of the rotating paddle wheels. The paddle wheels, in turn, lose this energy by friction with the water in the insulated container. If care is taken to minimize other frictional energy losses, one may postulate that the energy going into the microstructure of the water can be directly equated to the potential energy change of the falling mass. The energy (in calories) absorbed

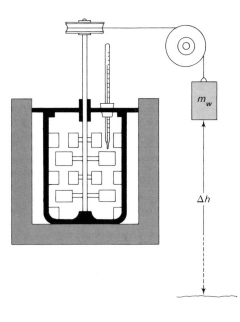

figure 7-8   A device for measuring the quantitative relationship between heat and mechanical energy. (From *Principles of Physical Science*, Second Edition, by F. T. Bonner, M. Phillips, and J. Raymond. 1971. Addison-Wesley.)

by the water is proportional to the product of the observed temperature change, $\Delta T$, and the mass of the water, $m_w$.

$$Mgh = cm_w \Delta T$$

Using units of meters, kilograms, and seconds for the quantities on the left-hand side of this equation results in an energy in joules, whereas substituting the value of $m_w$ in grams and $\Delta T$ in C° yields heat energy in calories, if the constant, $c$, has the numerical value of 1 for water. As a result of this and many other experimental arrangements for producing a measurable amount of internal energy by performing a known amount of mechanical work, Joule was able to show that a quantitative relationship exists between the joule and the calorie. The presently accepted value of this "mechanical equivalent of heat" is: 1 calorie = 4.185 joules.

Despite the demonstration of this equivalence, the practice of expressing heat energy, or internal energy, in calories and kinetic and potential energy in joules has not been altered. Note that this experiment does not *prove* that heat is another form of energy. Instead you *assume* that the potential energy has turned into internal energy in the microstructure, and then measure the relationship between calories and joules. All subsequent experiments using this relationship produce consistent results. It is this feature of general consistency that justifies an energy conservation law.

A German physicist, Hermann von Helmholtz, is primarily responsible for expanding the concept of the equivalence of mechanical and heat energies. The result is a more general energy conservation law that includes electrical, magnetic, chemical, and all other forms of energy. This law can be expressed in many ways, but in essence it denies the possibility of destroying or creating energy. In any process or interaction, the loss of one form of energy will result in a quantitatively equivalent increase in some other form of energy.

About fifty years before Helmholtz' work on energy conservation, the results of Antoine Lavoisier and others had led to the confirmation of another conservation law. This states the impossibility of creating or destroying the mass of a system. In his Special Theory of Relativity published in 1905, Albert Einstein combined the conservation of energy and conservation of mass laws. He postulated that mass and energy are just two different measures of the same thing. The formula $E = mc^2$ is a quantitative expression of Einstein's mass-energy relationship. $E$ is the energy equivalent to a mass, $m$, and $c$ is the speed of light ($3.0 \times 10^8$ m/sec). Using the formula we find that one kilogram of mass is equivalent to $9 \times 10^{16}$ joules. That is sufficient energy to operate a 100 watt electric light bulb for almost 30 million years, or to vaporize about 10 billion gallons of water! Thus, according to Einstein's theory, the apparent local loss of a small amount of mass should result in the release of vast quantities of energy. This postulate has been amply verified. The awesome result of a nuclear explosion is a dramatic example of huge amounts of energy in the form of heat, light, and sound produced at the expense of a modest amount of mass that had been locked in the solid material. The mass is not really lost but merely dissipated. The highly concentrated form of energy that we call matter becomes more widely distributed in the form of radiation which has a corresponding equivalent mass. Within the whole system, the mass remains the same. The seemingly endless production of heat and light by the sun and other stars is a result of continuous conversion of the concentrated mass of matter into the distributed mass of radiation.

Question 7.9    What is the mass equivalent of one calorie of heat energy?

The mass changes in the remaining products of most interactions in physics and chemistry are too small to be measured with weighing instruments. It is for this reason that the mass and energy conservation laws had been separately verified and accepted before Einstein pointed out that they were not independent. The conservation of mass-energy is one of the most important generalizations of physics. It has led to an understanding of a great many phenomena in all branches of science from astronomy to zoology.

The water at the bottom of a 60 meter waterfall is found to be warmer than the water at the top. What is the maximum value of the temperature difference?

Each kilogram of water will lose: $mgh = (1 \text{ kg})(9.8 \text{ m/sec}^2)(60 \text{ m}) = 588$ joules of potential energy. This potential energy will appear in the form of the kinetic energy of the falling water, which will then be converted to heat energy when the water is stopped abruptly at the base of the waterfall. The amount of internal energy produced will be: 588 joules/(4.185 joules/cal) = 141 cal. If one calorie will increase the temperature of one gram of water by one degree Celsius, 141 calories will increase the temperature of 1,000 grams of water by $141/1000 = 0.14°C$. The maximum value for the temperature difference, $\Delta T = 0.14°C$.

Joule was so preoccupied with the heat-mechanical energy conversion experiments that he took time off during his honeymoon in Switzerland to check his predictions on this waterfall effect. See Figure 7-9.

Question 7.10    The actual value of the difference in temperature of water at the top and bottom of a waterfall turns out to be smaller than the theoretical value calculated above. What assumptions were made that could account for this descrepancy?

## 7.4   effects of internal energy

Now that we have operational methods for defining temperature and heat, and units in which to express the quantities, let us see what happens when kinetic and potential energy disappear. From what we have just seen about

figure 7-9   Water in a waterfall will be slightly colder at the top or while it falls than at the base where some of its motion energy is converted to heat.

the mechanical equivalent of heat, we can arrange for this energy to be fed into material in a variety of equivalent ways. For instance, we could start with mechanical energy and arrange insulation around the apparatus so that as the mechanical energy decreases, all the internal energy that is produced is trapped. This is what Joule did. We could also start out with such heat producing devices as gas flames or electric coils. As long as these heat sources remain constant they can be calibrated in terms of their effect on a water bath. By definition, if an electric immersion heater raises the temperature of

100 grams of water 1°C every second, then it must be generating heat at a rate of 100 calories per second. Another method of adding a known amount of heat to material is to mix together two substances at different temperatures. This latter method is described in section 3 of Handling the Phenomena at the end of the chapter.

When energy goes into the microstructure of a material, it might cause changes of volume, pressure, state, temperature, or some combination of these properties. The amount of energy needed to change the temperature by 1°C depends on the amount of material present, the type of material, and the temperature. For water, this quantity is, by definition, one calorie per gram per Celsius degree at 15°C. (At other temperatures for liquid water this quantity varies by almost 1%.) For almost all other material it takes *less* than one calorie to raise the temperature of one gram by one degree. The measure of this quantity is called *specific heat*. The values for many different substances are given in the first column of Table 7.1.

*Question 7.11*   Note that the specific heat of water is 1. Is this a coincidence?

table 7.1   specific heats of some metals at room temperature

| Substance | cal/°C·g | (g/cm³) Density | cal/°C·cm³ | Mass in g/atom | cal/°C·atom |
|---|---|---|---|---|---|
| lithium | 0.81 | 0.53 | 0.43 | $1.2 \times 10^{-23}$ | $0.94 \times 10^{-23}$ |
| aluminum | 0.217 | 2.7 | 0.59 | $4.5 \times 10^{-23}$ | $0.97 \times 10^{-23}$ |
| copper | 0.092 | 8.9 | 0.82 | $10.1 \times 10^{-23}$ | $0.97 \times 10^{-23}$ |
| silver | 0.056 | 10.5 | 0.59 | $18.0 \times 10^{-23}$ | $1.01 \times 10^{-23}$ |
| tungsten | 0.032 | 19.3 | 0.62 | $30.6 \times 10^{-23}$ | $0.99 \times 10^{-23}$ |
| gold | 0.031 | 19.3 | 0.60 | $32.8 \times 10^{-23}$ | $1.02 \times 10^{-23}$ |
| lead | 0.030 | 11.3 | 0.34 | $34.5 \times 10^{-23}$ | $1.05 \times 10^{-23}$ |
| uranium | 0.028 | 19.0 | 0.53 | $39.6 \times 10^{-23}$ | $1.11 \times 10^{-23}$ |

As you can see from the first column in Table 7.1, there is a pattern in the values for the specific heats of solid metals. In general, the more dense the metal, the smaller the specific heat. There is a way to make this pattern more evident. Instead of calculating the energy required to raise the temperature of one *gram* of material, it might be just as reasonable to find the value per *unit volume*. For solids, surprisingly enough, this calculation:

$$\left(\frac{\text{cal}}{°C \cdot g}\right)\left(\frac{g}{\text{cm}^3}\right) = \frac{\text{cal}}{°C \cdot \text{cm}^3},$$

yields numbers (shown in the third column) that are close to each other.

In order to observe an even more remarkable regularity in the specific heats of metals we must anticipate some of the detailed information about the microstructure of matter that will be examined in the remaining chapters of this book. The changes caused in matter by an uptake in heat energy were attributed by Newton as early as 1704 to rearrangements and increases in agitation of an invisible submicroscopic structure of matter. As we shall see, our present understanding is that all matter is composed of basic building blocks called atoms. If we express our specific heats as calories per degree per atom by multiplying corresponding values in the first and fourth columns together (don't worry for the time being about how we have obtained the values given in column 4) the result is the set of numbers in the last column of Table 7.1. Assuming that our tiny atoms exist, and have the masses shown in column 4, it appears that for materials of the same type (solid metals, for instance), it takes the same amount of energy *per atom* to produce a one degree change of temperature. In other words, if you could take two pieces of aluminum and gold having equal numbers of atoms, and put equal amounts of energy into each chunk of metal, the temperature rise in each would be the same. The piece of gold would, of course, be a lot heavier than the aluminum, since each atom of gold is heavier than an atom of aluminum.

*Question 7.12*   We have mentioned three patterns that appear in specific heat calculations, depending on whether the value is given as: calories/degree · *gram;* or, calories/degree · *cubic centimeter;* or, calories/degree · *atom.* What are these patterns and how are they related? Since the specific heats of metals per unit volume are all about the same, what is implied about the volumes of the atoms involved?

Although there are many patterns and regularities in the values of specific heat for various substances, there are also some strange complications. For one thing, the specific heats do depend on temperature. Over a range of a few tens of degrees the specific heat of a substance is nearly constant, but important changes show up over larger ranges. Two examples of these changes are shown in Fig. 7-10. Notice that the specific heat values for solids are very different at very low temperature but approach the same value at very high temperature. In the case of hydrogen, the specific heat seems to change at certain particular temperatures. All of these regularities and complications are clues that can reveal information about the microstructure of matter. The exceptions are often as revealing as the rules.

Most materials expand when heat is fed into them. This expansion is the basis for the bimetallic thermometer described earlier. It is also responsible for the necessity of having expansion cracks in concrete paving. The whole slab of concrete expands when it gets hot, and if there is no room into which it can move the road will buckle. Not all substances behave in this way,

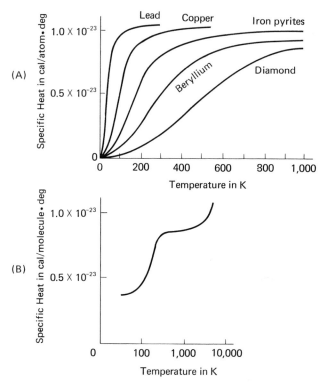

figure 7-10    (A) Specific heat (at constant volume) of hydrogen expressed in cal/(molecule)(degree) as a function of temperature. Note that the temperature scale is logarithmic. (B) Specific heat of various solids. Most metals (however, note the behavior of beryllium) have about the same specific heat at room temperature (300 K) when expressed in cal/(atom)(degree).

however. If energy is fed into solid water (ice), or bismuth, when they are at their respective melting points, the total volume of the resulting liquid will actually be less than the volume of the solid. Therefore, the density of the liquid is more than that of the solid. For this reason ice floats in water and solid bismuth floats in molten bismuth. As energy is fed into water at 0°C, the volume of the water decreases until the temperature has reached about 4°C. Above that temperature water acts normally and expands when heat is added.

If heat is fed into a gas kept at constant pressure the volume of the gas will always increase. While the fractional volume increase of solids and liquids is very small, that of gases is very large. The rules for this are described in Chapter 10, but for comparison note that a 1C° rise in temperature of mercury corresponds to a volume increase of only about 0.01%. The same temperature rise in a gas at constant pressure would produce a volume change of about 0.4%, a fractional change about 40 times as large as that in mercury.

Usually when heat is added to material the temperature rises. There are two reasons why this might not happen, however. The first is that the incoming heat might be turned completely into useful external work. For instance, as heat goes into a cylinder filled with steam, the gas could expand while pushing against a piston. Under the right conditions all of the incoming heat can be used up performing work and the gas could still be at the original temperature, though at a lower pressure. As a matter of fact, if the steam is at a high pressure to begin with, it can perform work by shoving the piston out without any heat being supplied. In this case the energy must have been stored in the gas. The loss of the internal stored energy is reflected by a drop in the temperature of the gas.

The second way in which heat can be absorbed in a material without any change of temperature is when the material is undergoing a change of state. It takes 80 calories to turn one gram of ice into one gram of water. While all this energy is entering the material the temperature remains at 0°C. If another 100 calories is added to the gram of water its temperature will steadily rise by 100 degrees. At that point it takes 540 calories to turn the water into steam, with the temperature remaining at 100°C.

Evidently heat energy can enter material and pass right out in the form of external work; it can be stored in the material without affecting the temperature but somehow influencing the physical state; it can divide up so that some of it changes the volume or pressure of the material and some of it is manifested as higher temperature. The time has come to examine the many facts about the effects of heat energy on matter in terms of a proposed model involving an atomic microstructure.

## 7.5   a model for the microstructure of matter

According to the model that will be developed in the next chapter all material, whether solid, liquid, or gas, is composed of tiny subunits called atoms. In many cases several of these atoms can combine in fairly stable combinations called molecules. Under certain conditions the molecules act as the major subunits and we do not have to consider the fact that they are composed of yet smaller parts. How small must we make the atoms so that our model will agree with experimental observations? Surely they must be smaller than we can see by eye or even with a microscope. The most powerful microscope does not reveal any granular atomic structure in matter. The limit of observability of any microscope using visible light is about one micron—one millionth of a meter ($10^{-6}$ m). That's about ten times smaller than the closest measurements made in most machine shops. The atom must be smaller yet. You yourself can actually make a measurement that lowers the acceptable upper limit for atomic size by another factor of 1000. That is, you can show that the atom must be smaller than $10^{-9}$ meters. The directions are given in Section 4 of Handling the Phenomena, at the end of the chapter. With that measurement, however, you will have come very close to the size of a single atom. Indeed, you will be measuring the length of a thin molecular chain about 20 atoms long. Any atom is about $2 \times 10^{-10}$ meters in diameter. This means that if you lined up one hundred million of them ($10^8$), shoulder to shoulder, they would fit along a line that is about two centimeters long.

*Question 7.13*    How many atoms would be in a solid cube, two centimeters on each side?

Our model for a gas cannot be one with the atoms right next to each other. We must assume that in a gas the atoms are separated from each other by considerable distances and are dashing around, bouncing off each other and the walls of the container. This model will be elaborated in Chapter 10, but certain crude features are required from what we have already seen about heat. It must be that as heat is fed into a gas the energy goes into the kinetic energy of the atoms. They bounce around faster and make more (and harder) collisions with the walls. As will be shown in Chapter 10, the pressure exerted by the gas is proportional to the average kinetic energy of the atoms. The physical property that we call the Kelvin temperature is proportional to the average kinetic energy of the atoms in any sample of matter.

*Question 7.14*    If this model is a good one, all of the energy supplied to a gas should be divided up among the atoms. At any given moment, some will be going fast (and some slow) but their average kinetic energy is proportional to the temperature of the gas. If equal amounts of heat are fed into two samples of different kinds of gas, *but each having the same number of molecules*, then the temperature rise of each sample should be the same. Therefore, the specific heats per molecule of any two gases should be the same. Check Table 7-2 and see if this is true.

Our model was too simple. The monatomic gases seem to act as we suggested, but apparently it takes extra energy to raise the temperature of diatomic gases. Here the subunits are molecules composed of two atoms held together. Such a system can absorb energy not only by moving around faster (translational motion), but also by vibrating and rotating. See Fig. 7-11. Since temperature is a measure of the average translational kinetic energy of each molecular system, the specific heat must be greater than it would be for single atoms. When heat energy is fed into such a gas, part of it goes into kinetic energy of the whole molecule, which makes the temperature higher, but part

table 7.2   specific heats of some gases at room temperature
and at constant volume

| Type of Gas | Gas | cal/°C·g | Mass in g/molecule | cal/°C·molecule |
|---|---|---|---|---|
| monatomic | helium | 0.74 | $0.67 \times 10^{-23}$ | $0.50 \times 10^{-23}$ |
| | argon | 0.075 | $6.7 \times 10^{-23}$ | $0.50 \times 10^{-23}$ |
| diatomic | hydrogen | 2.44 | $0.33 \times 10^{-23}$ | $0.81 \times 10^{-23}$ |
| | oxygen | .16 | $5.3 \times 10^{-23}$ | $0.84 \times 10^{-23}$ |
| | nitrogen | .18 | $4.7 \times 10^{-23}$ | $0.83 \times 10^{-23}$ |
| | chlorine | .09 | $11.8 \times 10^{-23}$ | $1.02 \times 10^{-23}$ |
| polyatomic | carbon dioxide | .15 | $7.3 \times 10^{-23}$ | $1.13 \times 10^{-23}$ |
| | sulfur dioxide | .12 | $10.6 \times 10^{-23}$ | $1.25 \times 10^{-23}$ |
| | ammonia | .39 | $2.8 \times 10^{-23}$ | $1.11 \times 10^{-23}$ |
| | ethane | .34 | $5.0 \times 10^{-23}$ | $1.71 \times 10^{-23}$ |

of it goes into the internal motions of the molecules. We might expect that
gases composed of larger and more complicated molecules would have even
higher specific heats, and as the table shows, this is indeed the case.

Another interesting feature about the specific heats of multiatomic gases
is that their specific heats increase with temperature in a strange manner.
A graph of the specific heat of hydrogen as a function of temperature was
shown in Fig. 7-10. Note that the specific heat appears to stay constant over
a considerable temperature range, then rises to a new value and stays constant
for another interval, and then rises again. It acts as if there were "threshold"
phenomena—something seems to happen that can't take place until the energy
gets above a critical level. At very low temperatures the specific heat of
hydrogen is the same as that of monatomic gases. Apparently the internal
structure of the molecule cannot absorb any energy. As the temperature rises

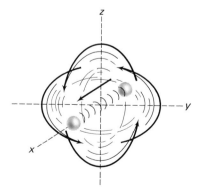

figure 7-11   A diatomic molecule can absorb energy in vibration and rotation.

above certain critical levels, various types of internal motion—vibrations and rotations—develop, which help to absorb the incoming heat energy. Such behavior could not be explained in terms of classical physics and is one of the phenomena that required the development of quantum theory. Some of the basic features of this theory are explained in Chapter 8. It turns out that the simple thermometer can actually probe into the microstructure of matter!

In Chapter 10 we take a careful look at the difference between gases and condensed phases (liquids and solids). For the time being it suffices to note that, in liquids and solids, the atoms and molecules are essentially touching one another.

The atoms are not free to wander around in a solid (though in some cases such migration can take place slowly). In our model we can picture each one surrounded by its neighbors, locked in place by electromagnetic forces. If two atoms get too close together they repel each other, and if they get too far apart they attract. Each atom is in a potential well whose general properties must resemble those in Fig. 6-16 on page 150. Under such conditions an atom will not remain stationary at the equilibrium position. Instead it will oscillate about that point. Heat energy fed into the material can be distributed among the atoms in terms of their energy of oscillation. Temperature in a solid must correspond to the vibration energy, since the atoms cannot dash around as in a gas. Our model yields a reasonable explanation of heat conductivity in a solid; vibrations at any one point of the material would influence the adjoining atoms, and so the energy would spread.

The apparent flow of heat from a hot object to a cold one through conduction can be visualized in terms of this model by examining the region of contact between the objects (Fig. 7-12). The atoms in the hotter object will be vibrating initially with greater energy (larger amplitude) than those in the colder object. Oscillation of any atom induces oscillations in its neighbors. In this way the energy will be transferred and eventually the two objects will come to the same temperature.

The internal energy of a solid consists of more than just kinetic energy of oscillation, however. An equal amount of energy (for a parabolic well) is

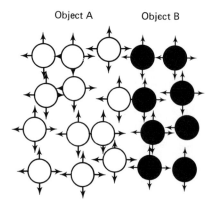

Object A        Object B

figure 7-12  A kinetic picture of "heat flow". When two objects are in contact, the molecules of the hotter object (open circles) transfer energy to the molecules of the colder object (filled circles) via intermolecular collisions.

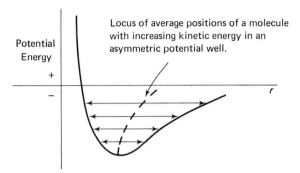

figure 7-13   The average position of an atom in an
asymmetric potential well changes with the kinetic
(vibration) energy, and hence with temperature.

in the form of potential energy of binding. As heat energy is fed into a solid
corresponding to our model, half of the energy would go into kinetic energy
of vibration of the individual atoms, and half would go into the increase of
potential binding energy. Note how easily this model explains the general
pattern of specific heat values for metallic solids. The value of
$1 \times 10^{-23}$ cal/°C · molecule is just twice that for monatomic gases. When we
provide $1 \times 10^{-23}$ cal to an atom of a metal, we assign half of it to increasing
the potential energy of the bound atom and the other half to increasing the
kinetic energy of oscillation. This latter energy (when all atoms in the object
have had the same treatment) manifests itself as a 1C° rise in temperature.

Our atomic model can explain other heat phenomena, at least qualitatively.
Note, in the schematic diagram of Fig. 7-13, that each atom is bound in a
potential well that is *asymmetric*. As the oscillations get larger, the atoms
do not continue to vibrate about their original equilibrium position. As Fig.
7-13 shows, the average spacing increases as the magnitude of vibration
increases. This slight increase of spacing between any two atoms explains
that fact that most materials expand as their temperature increases. Such
expansion is *not* explained in terms of the larger swings of the oscillating atoms.
So long as the equilibrium spacing does not change, the size of the material
will not change. Only the asymmetry of the potential well produces the
observed expansion.

What about the fact that ice and bismuth actually decrease in volume as
they melt and continue this perverse behavior over a small temperature range
above their melting points? In our model we must assume that in this region
the atoms or molecules are rearranging themselves in new configurations with
different potential wells. During such transitions internal energy might be
absorbed, or released, depending on the initial and final arrangements. At the
melting point of a solid, extra heat does not increase the amplitude of oscillation of the atoms, but instead provides the internal energy needed for breaking
some of the bonds that were holding the atoms or molecules in the locked

solid form. For most substances the liquid configuration has slightly more empty space and is therefore less dense than the solid. This need not be true and indeed the solid structures of bismuth and water—although more rigid— are more "open" than the liquids. In a liquid, any individual atom or molecule is still affected by its neighbors and oscillates within the potential well created by that effect. The well is relatively shallow, however, since it is not produced by the combined effect of many distant atoms. For any particular molecule the near-neighbor arrangement is temporary since the molecules can now migrate easily past one another. At the boiling point of a liquid the heat energy goes into freeing molecules from their binding to the other molecules. The average kinetic energy of oscillation—the temperature—of the remaining molecules does not change during this process.

## 7.6   heat energy and probability

Our model of heat and the microstructure of matter leads to a strange conse- quence when we consider what happens to the mechanical energy lost by friction. Before it was lost, this energy was in the form of the coherent motion of an object. For example, in a block of wood sliding to rest on a board, all of its atoms were taking part in the general forward motion. After the block comes to rest there is no more organized motion. Instead, according to our model, the surface atoms of both the block and the board are now vibrating faster around their fixed average positions. The sum of all this extra vibrational kinetic energy represents only part of the original kinetic energy of the block. The rest has gone into potential energy of the bound atoms. Instead of the energy being in the form of a coherent motion of all the atoms, it has now turned into a chaotic arrangement of individual stretchings and vibrations concentrated among the surface atoms. From this original region of excitation the energy will gradually dissipate, spreading out to the surrounding atoms of the block, and board, and into the air.

There is seldom any problem in turning mechanical or electrical energy into heat or the internal energy of a material. Indeed, it is hard to transform mechanical or electrical energy into each other *without* part of the original energy being dissipated as heat. Friction is everywhere. The transformation of these other forms of energy to heat or internal energy, can be 100% efficient. However, to turn the internal energy of the microstructure back into mechan- ical or electrical energy is another matter. Of course, it can be done. Most of our civilization gets its energy that way. All engines with cylinders make use of the internal energy of hot or exploding gases to push the pistons. The generators in most of our electric power plants are powered by steam that has been produced by heat energy derived from burning coal or oil, or from nuclear transformations. How is it that the chaotic motions of individual atoms vibrating in potential wells or dashing around in a gas can be transformed into the coherent motion of a piston or turbine?

*Question 7.15*   Consider a gas at higher than atmospheric pressure in a closed cylinder with a circular piston at one end. There are just as many atoms or molecules going one way as another. There is a distribution of velocities, centered on some average velocity that depends on the temperature. Explain how any coherent motion can be extracted from this chaos in order to yield useful mechanical work.

 

    This process of turning internal energy of the microstructure back into other forms of energy cannot be 100% efficient. Consider the situation in terms of probability. Left to itself, a complicated system will always degenerate into the most probable arrangement, assuming that it is free to transform from one arrangement into another. If you place 100 pennies with their heads up in a container and shake them thoroughly, it is highly unlikely that you will end up with 100 heads (or 100 tails) facing up. Instead, you will get a mixture of heads and tails. There is only one combination of the hundred coins that will give all heads, but there are many ways of getting combinations that show a nearly equal amount of heads and tails. Given a chance, the system will move toward more probable arrangements. Similarly, if you have a bunch of atoms all moving in the same direction (because the whole object is moving), and then there is a chance for that energy to be shared among other directions, you will end up with the more probable distribution of chaotic motions. It would be unlikely, however, for all of these unrelated motions to gather themselves together again and become a coherent motion of the whole object. Energy could be conserved if a hot block of wood were suddenly to cool off and go shooting across a table, but who would believe that such an unlikely phenomenon could occur?

    Nevertheless, heat engines do extract coherent motion of large objects out of the chaotic motion of individual molecules. The system moves from a more probable arrangement to a less probable one. We pay for it, however. To make such a process work, some heat energy has to be thrown away at a lower temperature. To see why this must be, consider the gas in the cylinder of a heat engine, as shown in Fig. 7-14. The general idea behind the functioning of this heat engine is that the heat admitted at one stage of a four step cycle will make the gas expand. The expanding gas moves the piston, thereby transforming the heat into work. However, in order to restore the cylinder to its original condition, the machine must force the piston back. Work is done in accomplishing this, and this work produces more heat. (You might have experienced this effect while pumping up a tire with a hand pump.) This heat must be given off, or it will raise the temperature of the gas. Now let us analyse the process, step by step.

    During Step 1, heat enters the gas but no temperature change results. The reason that the gas can remain at the same temperature is that it is expanding, and driving out the piston. (In actual machines, the temperature might be changing too, but the simple 4-step process being described is an approxi-

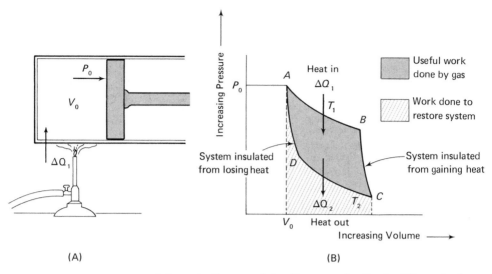

(A)                                    (B)

figure 7-14   Schematic diagram of the Carnot cycle. On the $PV$ graph, the path from $A$ to $B$, (line $AB$) at constant temperature $T_1$, is Step 1. The path $B$ to $C$ represents Step 2. The path $C$ to $D$, (line $CD$) at constant temperature $T_2$, is Step 3; and the path $D$ to $A$ is Step 4.

mation that is easier to analyze.) During Step 1, *all* of the heat energy is transformed into the mechanical work of shoving out the piston.

*Question 7.16*   This last statement appears to contradict our previous assertion that the efficiency of turning heat into mechanical energy had to be less than 100%. How can this be? (Consider what has happened to the gas and machine during Step 1.)

   If we wanted to restore the machine to its original condition, we could push back on the piston in such a way that heat would flow out of the cylinder, with the system remaining at the original constant temperature. On the pressure-volume graph of Fig. 7-14 this would be represented by retracing the path of Step 1 exactly backwards. This reversed process would take the same amount of work that was produced in the expansion, and the same amount of heat would be expelled that was originally absorbed. We would be right back where we started, with no heat turned into mechanical energy, and no heat lost. Such a process would not make a very useful heat engine!
   Instead, we go on to Step 2, in which the piston continues to move slowly outward although we do not allow any further heat to enter the system. The gas will expand in this way because it is hot and at higher pressure than the

surroundings. However, since no more heat energy is entering the system, the energy going into the piston must be coming out of the internal energy of the gas. Therefore the temperature of the gas will fall. Remember that in Step 1 the external heat that we supplied was turned into mechanical energy, and now in Step 2 some of the internal energy of the gas is turned into outside work. We are left, however, with an expanded cylinder and a gas having a lower pressure and temperature than when we started. In order to keep on using the engine, we must put it back to its original condition. Step 3 shows one way to do this. Pushing the piston in slowly, so that during this compression the gas temperature does not rise, necessarily drives out heat energy during the compression. Here is the crucial stage during which we have to release some heat at a lower temperature than that of Step 1.

At a particular volume, we then insulate the cylinder so that no more heat can flow out and keep on shoving in the piston. During this Step 4, the pressure and temperature will rise to their original values, as the volume decreases to its original value. The heat engine is now back in its starting condition, ready to repeat the cyclical process. During Step 1, an amount of heat, $Q_1$, was absorbed; during Step 3, a smaller amount of heat, $Q_2$, was expelled. During Steps 1 and 2, external work was performed; during Steps 3 and 4 we had to do work on the piston in order to restore the engine to its original condition. The amount of heat turned into mechanical energy was $(Q_1 - Q_2)$.

*Question 7.17*   During Step 1, all of $Q_1$ was turned into work; during Step 3 all of the work that we did on the piston turned into heat energy $Q_2$ which was expelled. What happened to the work done during Steps 2 and 4?

Using the hypothetical 4 steps of our ideal heat engine, the absorption of heat energy $Q_1$ produced mechanical energy equal to $(Q_1 - Q_2)$. The efficiency of the operation is therefore: $(Q_1 - Q_2)/Q_1$. It turns out that this ratio is exactly equal to a ratio composed of the temperatures of Steps 1 and 3: $(Q_1 - Q_2)/Q_1 = (T_1 - T_2)/T_1$. These temperatures must be specified in the absolute, or Kelvin, scale. Indeed, the Kelvin scale is *defined* in terms of this process.

Some general comments are in order about this idealized heat engine. First of all, it is not so unworldly or farfetched as may appear. Some actual heat engines have pressure-volume graphs that are very similar to the simple one assumed. For instance, in a standard car motor, gasoline vapor and air are injected and compressed in what amounts to Step 4. At that point they are ignited and expand in what amounts to a combination of Steps 1 and 2. The engine then expels them and compresses the new gas in a combination of Steps 3 and 4. Despite the differences, note that heat (produced from chemical energy) does mechanical work, but it is necessary to expel the depleted hot gases in order to keep the cyclical process going.

The ideal heat engine that we have been analyzing follows the Carnot cycle, first described by the French scientist N. L. Sadi Carnot. No other heat engine can be more efficient in transforming heat into mechanical energy.

Although we described a heat engine in terms of a cylinder filled with gas undergoing very specific and simple changes, the conclusion about the efficiency of the process applies to any heat engine. The *maximum* efficiency for the conversion of heat to mechanical (or electrical) work is always given by the ratio $(T_1 - T_2)/T_1$, where $T_1$ is the absolute temperature of the heat source and $T_2$ is the absolute temperature at which a fraction of the heat must be expelled. There has been increasing concern in recent years about thermal pollution, or a man-made increase in the temperatures of our rivers and lakes. The source of this problem is the phenomenon that we have just described. In our electricity generating plants the temperature $T_1$ is the temperature of the high pressure steam produced by coal or oil fires, or by nuclear fission. Because of the high pressures and the corrosive properties of superheated steam, this temperature is usually limited to 1100°F (600°C, or 870°K). In our calculation of efficiency we must use the Kelvin, or absolute, temperature. After the steam drives the turbines it must be condensed and returned to the boiler. Low final pressure is produced by this condensation of the steam into water. The power stations must pump in large volumes of cold water to extract the heat given off as the steam pressure drops, and as the steam condenses. The final temperature of the steam (which is at a pressure much below atmospheric) might only be 35°C (310°K).

*Question 7.18*   With an input temperature of 870°K and an exhaust temperature of 310°K, what is the maximum thermal efficiency of such a power station?

Actual efficiencies of modern generating plants are less than 50% because of various internal power losses. Under such conditions, for every unit of energy extracted from the fuel, less than half a unit is transformed into electrical energy and the rest literally goes down the drain.

The study of how heat can be transformed into mechanical energy is one of enormous practical importance and also of very profound theoretical interest. The subject is called *Thermodynamics,* and much of it is summarized in two profoundly significant laws. The first law is really an assertion that heat is a form of energy; and that in heat-mechanical transformations, energy is conserved. In symbolic terms, the first law can be stated as: $\Delta Q = \Delta U + \Delta W$. The heat entering a system, $\Delta Q$, is equal to the change in internal energy of the system, $\Delta U$, plus any external work performed, $\Delta W$. For instance, when heat is fed into a gas, the temperature of the gas may rise (indicating an increase in internal energy) and the gas may expand against a piston doing useful external work. The second law of thermodynamics is more subtle and can be expressed in several ways, all of which are equivalent although they may not seem so. One way of stating it is; if a system is free to transform itself in a variety of ways, it will end up in a state of higher probability. For instance, heat will always flow from a hot region to a colder region.

*Question 7.19*    Why should the direction of heat flow have anything to do with probability?

Another form of the second law makes the claim concerning heat engines that we have just investigated. It is impossible to convert heat into mechanical energy in a cyclic process without expelling a fraction of the heat at a lower temperature. Incidentally, a corollary of this law is that it takes work to drive heat from a low temperature system and expel it at a higher temperature. We all know this anyway. In a refrigerator, heat is taken from cold objects, making them colder, and is expelled into the warmer room air through the radiator. The unnatural direction of heat flow is maintained, however, only so long as the electric cord is plugged in so that the motor can supply the necessary work.

The implications of the second law are universal. All processes, here on earth and elsewhere, occur in such a manner that the overall system moves to more probable states. Disorder is more probable than order, however. The concentrated forms of energy, and orderly systems, inevitably degenerate into dissipated and chaotic arrangements. The physical property that refers to the relative state of chaos of a system is called *entropy*. No energy is lost; it is simply degraded into less available forms. We can defeat this trend locally in many ways. Living creatures organize materials from the more random arrangements around them. (A living cell is obviously more organized than the raw materials from which it is constructed.) Energy is required, however, for the production of this organization. That energy comes from the sun, where concentrated energy in the form of the mass of atomic nuclei is partially transformed into radiant energy. Overall, there has been a reduction in the amount of order. So far as we know, there is no escape from the second law of thermodynamics. Can you explain the caption to Fig. 7-15?

## summary

In this chapter we started out to find an answer to the following question. If we want to maintain the law of conservation of energy, how do we explain the disappearance of mechanical energy in most moving systems? The experimental observation is that when kinetic and potential energy disappear, objects change their state, or change volume, or get hotter. In order to make such observations quantitative, it is necessary to differentiate between heat and temperature and to define these terms operationally. Temperature was defined

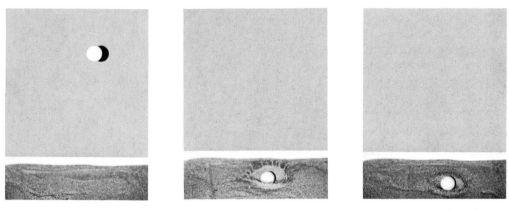

figure 7-15   Golf ball falling into a box of sand. The sequence occurred as shown, but could not happen backwards.

in terms of the behavior of certain materials that expand when heat is supplied to them. We described a method for calibrating thermometers and compared several standard scales. The calorie unit of heat is chosen as that quantity which will raise the temperature of one gram of water by 1C°. Over a century ago Joule showed experimentally that there was always a constant relationship between the amount of mechanical energy lost in a system and the equivalent internal energy provided. Heat is energy, but we still must explain the details of what happens to the energy inside material.

To answer that question we proposed a model of the microstructure of matter. Subunits, called atoms; or, atoms linked together as molecules, can exist in three main states. In gases they are widely separated, dash around at high speeds, and elastically bounce off each other and the walls of containers. Any heat supplied to a gas goes, at least partially, into the kinetic energy of the atoms. If the gas is composed of molecules, some of the energy can go into the internal vibrations and rotations of the molecules. Consequently, the specific heat of a gas composed of complex molecules is greater than that of a gas composed of single atoms.

The specific heats of materials show many uniformities if they are expressed in terms of calories/[molecule (or atom)] · (°C). In this form we measure the temperature increase when a unit of heat energy is supplied to *equal numbers* of the subunits. All metals have almost the same atomic specific heat, but the value is twice as great as that for monatomic gases. According to our model, solids are composed of atoms or molecules locked closely together in potential wells created by their neighbors. Heat energy is absorbed by the kinetic energy of vibrations of the bound atoms, and by their increase in potential energy. Each mode—kinetic and potential—gets about half of the absorbed energy.

Energy conservation includes mass conservation, since mass and energy units simply reflect different ways of measuring the same thing. The first law of thermodynamics asserts that heat is energy, and that energy is conserved in thermal processes. The second law of thermodynamics restricts the way in which energy transformations can take place. Systems must transform in the direction of the most probable arrangements. Consequently, even though kinetic and potential energy can be turned into heat with 100% efficiency, heat engines must reject some of their heat input at a lower temperature, in order to turn heat into work. This downgrading of energy is a universal phenomenon, brought about by the inevitable change from order to disorder.

*Answers to Questions*

7.1   In writing your answer, consider how you would measure each quantity.

7.2   Caloric fluid could be combined with the material substance of a body in such a way that it is not "free" to manifest itself in the form of higher temperature. Friction might set some of this heat fluid free. (In fact, explanations such as this were proposed.)

7.3   The volume of the bulb at the end of a thermometer is very large compared to the volume of the small diameter capillary tube to which it is connected. The increase in volume represented by a relatively large fractional increase in the length of the liquid column in the tube is thus still a very small percentage increase for the total liquid confined in the thermometer.

7.4   As the temperature is lowered the bimetallic strip will become flat and then bend the other way, since the metal that expands more per degree of temperature change, also contracts more when the bimetallic strip cools.

7.5   Solidification of a hard boiled egg, charring of a piece of wood, melting of a snowflake, etc.

7.6   The mercury will solidify at $-38°$ and boil at $356°$ on a temperature scale constructed in this manner.

7.7

7.8   Since there are $180°$ between the freezing and boiling points of water on the Fahrenheit scale and $100°$ between these same reference points on the Celsius scale, a change of $x$ Celsius degrees will correspond to $(180/100)x = (9/5)x$ Fahrenheit degrees. To convert Celsius to Fahrenheit we must, therefore, first multiply the Celsius reading by $9/5$. Since the melting point on the Fahrenheit scale is $32°$ instead of $0°$, we must add $32°$ to complete the conversion.

   Absolute zero is $-273.2°C$. To convert this to $°F$ we make use of the formula given in Question 7-8, thus:

$$T(°F) = 9/5 \, (-273.2) + 32$$

$$= -491.8 + 32 = -459.8°F$$

Absolute zero is $-459.8°F$

7.9   $E = 1$ calorie $= 4.18$ joules. Substituting this value for $E$ in the equation $E = mc^2$ yields:

$$4.18 \, J = m \, (3.0 \times 10^8)^2$$

$$= m \, 9.0 \times 10^{16}$$

$$m = \frac{4.18 \times 10^0}{9.0 \times 10^{16}} = 0.46 \times 10^{-16} = 4.6 \times 10^{-17} \, kg$$

7.10   The calculation assumed (1) that all of the kinetic energy of the falling water was converted to heat at the bottom of the waterfall and (2) that this energy was completely retained by the water. In an actual case, some of the kinetic energy might not be converted to heat, but will show up as a more rapid flow of water at the base of the waterfall. In addition, part of the heat produced in the water will be lost to the air and the river bed. Furthermore, some of the falling water will evaporate, producing a cooling effect on the remaining water.

7.11   No! The calorie was defined in such a way that the specific heat of water is unity.

7.12   For the specific heats of metals in cal/degree · gram, the pattern is one of decreasing values as the density increases. If each of these values is multiplied by the mass, in grams, of an atom of that metal, the expression becomes cal/degree · atom. The pattern for these values (Table 7-1 column 5) is that numerically they are almost all the same. In column one, the values are decreasing by 1/(atomic mass in grams) as shown in column four. Multiplying by the atomic mass should indeed produce a constant relationship. Since the values of specific heat in terms of cal/degree · cm³ are nearly the same for most metals, it must be that each metal has about the same number of atoms per cubic centimeter. Therefore, the diameters of such atoms must not vary greatly.

7.13   If there are $10^8$ atoms along each side of a cube, then there are $(10^8)^3 = 10^{24}$ atoms in the cube.

7.14   The molecular specific heats of both monatomic gases listed are the same, but, this value is different from those of the diatomic gases.

7.15   In a chaotic situation with motion in all directions, energy can be extracted if some energy absorbing device in the system can move in one direction but not in any other. The motion of the gas in a cylinder produces a pressure on the inside surface of the cylinder. If the only part of the system that can move is the piston, the random bombardment of the piston by the gas atoms will drive it outward, doing external work, and absorbing energy from the gas.

7.16    It is true that all of the heat energy in Step 1 is turned into mechanical energy, but at the expense of changing the state of the heat engine itself. At the end of Step 1 the piston is located at a different position than at the beginning. This process of turning heat into mechanical energy could not continue without getting the piston back to its original position so that the action could be repeated over and over again with same engine.

Note, incidentally, that the area under a curve on a graph of pressure *vs.* volume represents the work done. Work is obtained by multiplying force and distance through which the force is exerted. In the case of a cylinder and piston, the total force exerted is equal to the pressure, $P$, times the cross sectional area, $A$, of the piston: $F = PA$. The work, $W$, done by the piston is equal to the average force, $\bar{F}$, times the distance, $x$, it travels: $W = \bar{F}x = \bar{P}Ax = \bar{P}(\text{volume swept out})$. ($\bar{F}$ and $\bar{P}$ are the *average* force and pressure respectively.)

7.17    During Step 2 the internal energy of the gas produced external work and the temperature dropped. During Step 4 external work had to be done to restore the gas to its original condition, including raising the temperature. It turns out that the work obtained from the gas in Step 2 is just equal to the work that had to be performed on the gas in Step 4. There is no net gain or loss.

7.18    Thermal efficiency $= (870 - 310)/870 = 560/870 = 0.64$

7.19    Let us suppose that the bound atoms at one end of a metal rod were all vibrating with large amplitudes, while at the other end they were all vibrating with small amplitudes. In other words, one end of the rod is hotter than the other. Since vibrations in one region of the metal lattice start up vibrations in the neighboring regions, after a while all the atoms in the rod will be vibrating with the same amplitude, which will be somewhere inbetween the original large and small amplitudes. Heat will have flowed from a hot area to a cold one. It would be very improbable if all of the atoms at one end of a rod originally at a uniform temperature suddenly started vibrating more vigorously than those at the other end, even though the average energy in such an event could be conserved.

handling the phenomena

1.    *Calibration of a mercury thermometer*
Use the procedure described in the text and illustrated in Fig. 7-4 to calibrate an ungraduated mercury thermometer. (Ungraduated thermometers are available from several scientific supply companies.) You can use glass marking pencils to mark the zero and 100° reference points and to make intermediate markings in divisions of 10°. For the reference mixture at 0°C, ice cubes and water will *not* be satisfactory because it is too difficult to keep them properly mixed. Instead, use a mush of chopped ice and water contained in a styrofoam cup. (Check the actual boiling point of water at your altitude. In Denver, for instance, water boils at 95°C.)

## 2.    *Conversion of mechanical energy to heat*

The equipment needed for this experiment is a cardboard mailing tube about 1 meter long sealed by screw caps at both ends, several hundred grams of lead or copper shot, and a thermometer capable of 0.2°C precision.

Record the temperature of the shot. Carefully roll the shot down the mailing tube and seal the tube with the end caps. Invert the tube rapidly so that the shot falls from one end to the other, having as little contact with the walls of the tube as possible. Repeat the inversion of the tube 99 more times. Quickly insert the thermometer and measure the temperature of the shot. Calculate the number of calories generated by the shot in its total fall of 100 meters.

Determine the specific heat of the metal from the temperature increase that you observed assuming that no heat is lost during the experiment. Would the observed temperature change be different if a 400 gram sample were used instead of a 200 gram sample?

Discuss the sources of error in this experiment. Look up the specific heat of the metal used in Table 7-1 and see how it compares with your result. How could this experiment by improved in order to give a more accurate result?

## 3.    *Heat exchange in mixtures*

Heat can be introduced into an object through the use of flames, electric current, or by doing work on the object. Heat can also be exchanged between the components of a mixture. For instance, if hot water is added to cold water, the part of the water that was originally hot provides heat to the part that was colder, producing a mixture all at some intermediate temperature.

(A)   Pour 50 cm³ of boiling water into 100 cm³ of water at room temperature and find the resulting temperature. Use two nested styrofoam cups for the mixture; it provides good insulation, especially if a cap is placed over the top. Before actually making the measurement, predict the expected result.

$$\text{heat lost by hot water} \longrightarrow \text{heat gained by cold water}$$

Remember that 1 cm³ of water has a mass of 1 gram, and that the specific heat of water is 1 calorie/gram · °C.

(B) To find the specific heat of a metal, pour a known amount of boiling water onto a known amount of metal in an insulated cup, at room temperature. Find the resulting temperature. Once again:

$$\text{heat lost} \longrightarrow \text{heat gained}$$

$$c_{water} m_{water} (T_i - T_f) = c_{metal} m_{metal} (T_f - T_{room})$$

where $c$ is the specific heat in calories/g · °C,
$m$ is the mass in grams,
$T_i$ is the initial temperature of boiling water (100°C),
$T_{room}$ is the room temperature, which is the initial temperature of the metal, and
$T_f$ is the final temperature of the mixture.

Use small pieces of the metal (like metal shot), so that temperature equilibrium can be reached rapidly with only a little stirring.

Since the specific heat of water is much greater than that of the metal, use only enough water to cover the metal. Otherwise, there will not be much of a temperature change for the water.

4. *Determining a molecular size by using a thin film*
   A drop of oleic acid oil on the surface of still water ideally could spread out until the film is only one molecule thick. If you start out with a measured volume you can measure the area covered by the oil slick, thus finding the thickness of the film and the size of the molecules,

$$\text{original volume} = (\text{area of oil slick})(\text{thickness})$$

To estimate the expected result, assume that the molecules are $10^{-9}$ m long ($10^{-7}$ cm) and therefore that this is the thickness of the film. If you start with 1 cm³ of oil, the final area of an oil slick $10^{-7}$ cm thick is given by:

$$1 \text{ cm}^3 = (\text{area in cm}^2)(10^{-7} \text{ cm})$$

If the oil slick is assumed to be circular, then its area $= 10^7$ cm². Since the area equals $\pi r^2$, substituting $\frac{1}{2}d$ for the radius yields $\pi[\frac{1}{2}d]^2$, or $(\pi/4)d^2$.
   Solving for $d$, the diameter of the slick becomes:

$$d = \sqrt{\frac{4}{\pi} \, 10^7} = \sqrt{\frac{40}{\pi} \, 10^6} = \sqrt{\left(\frac{40}{\pi}\right)} \times 10^3 \approx 3.5 \times 10^3 \text{ cm}.$$

You would need a large swimming pool!
   To perform the experiment easily in the lab, the diameter of the oil slick should not be greater than about 30 cm. You can decrease the size of the drop by using a medicine dropper. (Find how many drops equal 1 cm³.) To decrease the amount of oil even further, dilute the oil with alcohol. Calculate in advance whether you should dilute the oleic acid by 10:1, 100:1, or by some other ratio. The alcohol will mix with the water and not produce a film of its own. Oleic acid is a long chain molecule ($C_{17}H_{33}COOH$) with the strange but useful property that one end is repelled by water and the other end is attracted to it. Hence the molecules in the surface film line up with their longest dimension in a vertical direction.

**CAUTION!** The most important experimental precaution is to avoid contaminating the water with other types of oil. Use a fresh aluminum foil surface on a cafeteria tray to create a suitable pool. Don't even touch the aluminum surface with your fingers, lest extra oil get into the water.
   An oil film this thin cannot ordinarily be seen. To make its presence visible, dust the surface of the water *very* lightly with lycopodium powder (or a fine talc), before the oil is added. The spreading oil will shove the dust ahead of it, leaving an irregular clear patch. Estimate roughly the area of the film, and calculate the thickness of the oil film, and hence, the length of the molecule.

problems

1  A mercury thermometer consisting of a spherical bulb having an inner diameter of 0.4 centimeter and a tube having an inner diameter of 0.01 centimeter is heated until it shows a 5.0C° rise in temperature. The volume of mercury expands by 0.18 percent for each 1.0C° increase in temperature. Calculate the increase in height

of the mercury that corresponds to this 5.0C° temperature rise. (Recall that the formulas for the volumes of a sphere and a cylinder are $\frac{4}{3}\pi r^3$ and $\pi r^2 h$, respectively.)

2   Examine an inexpensive oven thermometer. Explain how it works. (The metal coil in the back is a bimetallic strip.)

3   When two wires made of different metals are joined together at both ends, and the two junctions are at different temperatures, an electric current will flow through the resulting wire loop. The voltage difference between the two junctions—which can be measured by a sensitive voltmeter—will depend on their difference in temperature. Describe how you could construct a thermometer based on this experimental observation and how you would calibrate it. (Such devices, called thermocouple thermometers, are actually used quite often for a wide variety of technical applications.)

4   A recipe in a French cookbook specifices an oven temperature of 175°. What Fahrenheit setting should you use?

5   Normal body temperature is about 98.6°F. What is the equivalent temperature in °C?

6   If the falling mass in Joule's apparatus (Fig. 7-8) is 50 kilograms and it falls a distance of 200 centimeters what will be the resulting temperature change if the apparatus holds 500 grams of water? (List any assumptions used in arriving at your answer.)

7   In an atomic explosion some of the nuclear mass is distributed in the form of kinetic energy of the explosion particles. Assume that such an explosion occurs at the bottom of a small lake containing 100 million liters of water and that exactly 1.0 gram of mass is converted to heat energy that is contained by the water. If the lake is initially at 25°C, what will be its final temperature?

8   Explain why early printing type makers found that they could get much better results if they used a metal (or actually a solution of several metals called an alloy) which expanded when it solidified in their dies, rather than a metal like lead which contracts.

9   Assuming that the metal atoms have a diameter of about $2 \times 10^{-8}$ cm and are all touching, estimate the number of atoms in a penny.

10   When a steel ball is struck with a hammer, or heated with a match, the kinetic energy of its atoms is increased. Why does it move in one case but not the other?

11   Two engines both produce useful work with an efficiency of 75 percent. One engine has an exhaust temperature of 10°K and the other has an exhaust temperature of 300°K. What is the minimum input temperature for each engine?

12   In everyday life we see many examples of spontaneous processes which involve an increase in entropy. Give some examples of the natural tendency of things to degrade into a state of random chaos from an original state of order. In each case consider the energy required to reestablish the state of order.

# atoms and the periodic table

During the last two chapters we have found it necessary to propose a model for the internal structure of matter. We have found a satisfactory explanation for the concepts of heat and temperature in terms of submicroscopic particles which can soak up energy by moving and by stretching and compressing their mutual bonds. To gain a further understanding of this phenomenon as well as the variety of other interactions which involve changes in the structure of matter, let us investigate the nature of the microstructure in considerable detail.

Professor Richard P. Feynman, Nobel laureate and one of the world's outstanding physicists, has written:

"If, in some cataclysm, all of scientific knowledge were to be destroyed, and only one sentence passed on to the next generations of creatures, what statement would contain the most information in the fewest words? I believe it is the *atomic hypothesis* (or the atomic *fact*, or whatever you wish to call it) that *all things are made of atoms—little particles that move around in perpetual motion, attracting each other when they are a little distance apart, but repelling upon being squeezed into one another.*"[1]

Others may disagree on exactly what should be in that "one sentence," but something about the existence and nature of atoms would surely be high on any scientist's list.

---

1. *The Feynman Lectures on Physics*, Addison-Wesley Publishing Company, Inc. (1963), Reading, Massachusetts. Pg 1-2.

It may seem strange to someone who has been born into a world in which the word atom is commonly found in the vocabulary of preschool children that scientists would consider the existence of atoms an obscure enough fact to demand first place among the secrets of the universe that have thus far been unraveled. Although the notion that all matter is composed of ultimate submicroscopic particles dates back at least as far as the fifth century B.C., the fact is, that only in this century has the atomic theory gained virtually universal acceptance.

## 8.1   dalton's atomic theory

The credit for conceiving the basic model from which our present understanding of atomicity has grown goes to the English school teacher turned chemist, John Dalton. In 1808, he published a complete exposition of his theory in a book entitled *A New System of Chemical Philosophy*. The recognition accorded to Dalton is clearly not due to his being the first to suggest the existence of atoms—the Greek philosophers, Leucippus, Democritus, and Epicurus, and such famous Seventeenth and Eighteenth century scientists as Robert Boyle and Isaac Newton professed a belief in the atomicity of matter. It remained for Dalton to develop a *quantitative theory* around the atomic hypothesis which could be used to make *predictions* that could be *tested*. Dalton was not a particularly talented experimentalist (a fact which might be of some comfort to aspiring young scientists who discover that they are all thumbs in the laboratory), but he had a great gift for original thought, and the ability to distill unifying principles from an examination of the experimental results of others.

Prior to Dalton, considerable painstaking experimentation by Seventeenth and Eighteenth century scientists had provided compelling evidence that all matter is ultimately resolvable into a relatively small number of "simple" substances or *elements*. More than two dozen of these elements had been identified and characterized. Careful quantitative experiments championed by the French chemist, Antoine Lavoisier, led his countryman, Joseph Proust, (a contemporary of Dalton) to the conclusion that a pure *compound* always contains the same fixed ratios, by weight, of the elements of which it is composed. This statement, commonly referred to as the *law of definite proportion*, was challenged by a third French chemist, Claude Berthollet, whose own experiments indicated that the weight ratios of the elements in compounds were variable. He offered several apparent examples of variable composition as evidence. These included the compounds formed when certain metals reacted with oxygen under varying conditions of pressure and temperature. Proust was able to show that these supposedly pure compounds were, in fact, mixtures of varying proportions of two or more oxides of the same metal. (Fig. 8-1) For example, copper forms two oxides, one of which contains

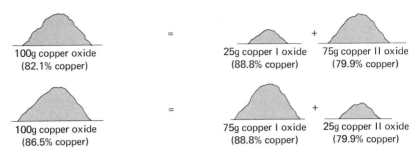

figure 8-1   An apparent variation in the percentage, by weight, of copper in different samples of copper oxide is illustrated. Variations such as these are due to the presence of different proportions of two unique oxides in the samples. Copper I oxide (red) contains 88.8% copper by weight, whereas Copper II oxide (black) contains 79.9% copper by weight.

79.9% and the other 88.8% copper by weight. Mixtures of these two unique oxides (in appropriate proportions) can produce samples having a copper content anywhere between these two limiting values.

*Question 8.1*   A sample of "copper oxide" is found to contain 83.1% copper, by weight. What are the proportions of the two unique oxides in this sample?

In developing his atomic theory, Dalton provided the basis for understanding the law of definite proportion. The major postulates in Dalton's treatise can be paraphrased as follows:

1   All matter is composed of ultimate indivisible particles called atoms.
2   Each different element corresponds to a distinct atom—there are as many different atoms as there are elements. The atoms of a particular element always have the same mass and other characteristic properties.
3   Atoms can never be changed, one into another.
4   Compounds have as their basic units groups of definite numbers of different atoms called "compound atoms" (molecules, in modern terms).
5   Chemical changes involve the rearrangement of atoms but never their creation or destruction.

*Question 8.2*   Explain how the law of definite proportions is explainable in terms of Dalton's theory.

Dalton gained support for his theory by using it to formulate, and then successfully verify, the *law of multiple proportions;* when more than one compound is formed from the same two elements, the ratio of weights of one element which will combine with a given weight of the second element is always equal to the ratio of small whole numbers. The atomic theory could be used to explain this rule in the following manner. Suppose Compound I has molecules containing three atoms of A for every two of B and Compound II has five atoms of A for every two of B. Then the ratio of *weights* of A combining with a *given weight* of B will be 5:3 for the two compounds. See Fig. 8-2. As an example, consider two compounds containing only hydrogen and oxygen—water ($H_2O$), and hydrogen peroxide ($H_2O_2$). For each gram of hydrogen produced in the decomposition of water, *eight* grams of oxygen will be produced. In the case of hydrogen peroxide, one gram of hydrogen will be associated with *sixteen* grams of oxygen. For the same weight of one element (one gram of hydrogen in each case), one compound yields 8 grams of oxygen and the other compound yields 16 grams of oxygen. The ratio 16:8 is a ratio of small whole numbers (2:1).

Note that the *law of definite proportions* refers to each compound separately. The ratio of weights of oxygen to hydrogen in (ordinary) water is always 8:1. For most compounds this ratio need not be a small number and usually is not a whole number. The *law of multiple proportions* is concerned with two or more compounds of two or more elements. The comparison of weights is always made by taking the same weight of one element in each compound and comparing the weights of another element in the two compounds. This ratio is always the ratio of small whole numbers. Both of these laws are most

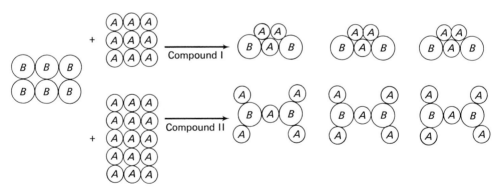

figure 8-2 *The law of multiple proportions.* Elements A and B form two compounds. Compound I molecules contain three A atoms and two B atoms each. Compound II molecules contain five A atoms and two B atoms each. If we take any specified weight of element B, the atomic theory requires that it contain a specific number of B atoms. To form compound I from this specific number of B atoms will require three A atoms for every two B atoms. To form compound II from the same specific number of B atoms will require five A atoms for every two B atoms. Since all A atoms have the same weight, the amounts of element A required to react with a specific weight of B to form compounds II and I must be in the ratio of the small whole numbers, 5:3.

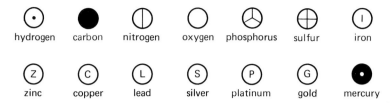

figure 8-3  Dalton's ideograms representing atoms of some of the elements.

easily explained by assuming that every chemical compound always consists of a specific combination of individual atoms.

Dalton developed picture symbols or ideograms to represent each atom. Some of these are shown in Fig. 8-3. Today we use single capital letters, or pairs of letters with the first capitalized, to designate each of the separate atoms. The letter symbols for the first eighteen elements are given in Table 8-1.

table 8.1    the symbols for the first eighteen elements.

| 1. Hydrogen | H | 10. Neon | Ne |
|---|---|---|---|
| 2. Helium | He | 11. Sodium | Na |
| 3. Lithium | Li | 12. Magnesium | Mg |
| 4. Beryllium | Be | 13. Aluminum | Al |
| 5. Boron | B | 14. Silicon | Si |
| 6. Carbon | C | 15. Phosphorus | P |
| 7. Nitrogen | N | 16. Sulfur | S |
| 8. Oxygen | O | 17. Chlorine | Cl |
| 9. Fluorine | F | 18. Argon | Ar |

## 8.2  classical models for the structure of the atom

By the beginning of this century, Dalton's notion of the indivisible atom had been challenged by the discovery that negatively charged particles called electrons could be obtained from many different elements. This suggested that electrons were subatomic particles, a notion that required the development of a model for the structure of the atom.

Since the electron is negatively charged and the atom is neutral, it was immediately recognized that atoms must also contain positive charges. J. J. Thomson, an English physicist, proposed a sort of raisin pudding structure with negative electrons embedded in a positively charged fluid as illustrated in Fig. 8-4.

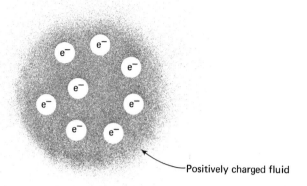

figure 8-4 "Raisin pudding" model of the atom. The fluid
carries just enough positive charge to neutralize the negative charge of the embedded electrons.

Between 1909 and 1911, a series of experiments performed in the laboratory
of Ernest Rutherford, in Manchester, England, clearly demonstrated that an
atom actually contains a positively charged nuclear core. Rutherford allowed
alpha particles to strike a very thin gold foil, as shown in Fig. 8-5. (Alpha
particles are spontaneously emitted by certain elements and were known, at
the time, to be positively charged, and about as massive as helium atoms.)

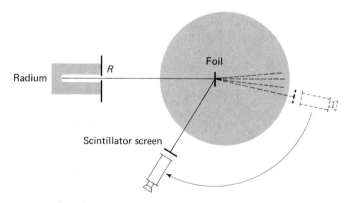

figure 8-5 Schematic diagram of Rutherford's experiment.
Alpha particles are emitted at *R*, scattered by gold foil and
observed by viewing their scintillations with a microscope focused on a movable screen, shown here in two positions. Most
alpha particles undergo very small deviations but a few are
scattered through large angles.

Calculations involving momentum conservation showed that the paths of these particles should be only slightly changed as they passed through an atom if its mass were spread out like a raisin pudding. Instead it was found that some of the alpha particles underwent large-angle scattering, and a few actually rebounded backwards. To quote Rutherford, "It was quite the most incredible event that has ever happened to me in my entire life. It was almost as incredible as if you fired a 15-inch shell at a piece of tissue paper and it came back and hit you." Rutherford correctly concluded that this result could occur only if all of the positive charges and most of the mass of the atom were concentrated in a very tiny part of the entire atomic volume. The ratio of the radius of this tiny nucleus to typical distances between adjacent atoms in a solid is about $1:30,000$.

*Question 8.3*  Explain why a nuclear atom would be expected to cause the results observed by Rutherford; whereas, the raisin pudding model would not.

The relatively large volume outside the nucleus must be occupied by electrons equal in number to the number of positive charges on the nucleus. It was assumed by Rutherford and his followers that, since the electrons had little mass compared to the nucleus, they probably were much smaller, and occupied even less space. This leads to a model of an atom consisting mostly of empty space! See Fig. 8-6.

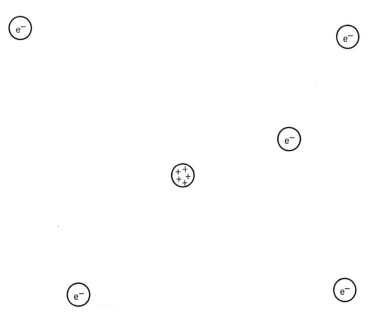

figure 8-6  Rutherford's conclusion—that most of the mass of the atom was concentrated in a tiny positive nuclear core—led to the realization that most of the volume of the atom was empty space.

Rutherford's results raised several new questions about the structure of the atom. Scientists began speculating about the way in which the electrons were distributed in space around the nucleus, and what prevented the attraction between the positive nucleus and negative electrons from causing the atom to collapse. An obvious suggestion is that the atom is analogous to the solar system. Unfortunately, the classical mechanical model of a positively charged nucleus encircled by orbiting negatively charged electrons does not result in a stable situation such as we examined in Chapter 6 for gravitational satellite systems. According to the well established principles of electricity and magnetism, a charged particle which changes either its direction or speed will radiate energy in the form of electromagnetic (light) waves.

*Question 8.4*    Why does the prediction that energy would be radiated mean that the solar system model of the atom would be unstable?

## 8.3    the "new physics" and the modern model for the atom

The first major step toward the development of the presently accepted model for the atom was taken by Niels Bohr, a Danish scientist who went to work in Rutherford's laboratory at the time the experimental proof of the nuclear atom was being published. Before Bohr's contribution is described we must take a brief look at some other developments that were to result in sweeping changes in the understanding of all submicroscopic phenomena.

The theory of the atomic nature of matter, and the notion that mass was not indefinitely divisible, had gained wide (but not complete) acceptance by the end of the 19th century. However, it was still assumed that energy and momentum were transferable in *any* amount, no matter how infinitesimal. In the year 1900, in an attempt to understand the properties of electromagnetic radiation, the German physicist Max Planck proposed that this form of energy was emitted in a discontinuous or "quantized" manner. Electromagnetic radiation is the general phenomenon of which visible light is a particular example. When any object is heated it emits energy in the form of electromagnetic waves, as shown in Fig. 8-7. The distance between the peaks in the waves is called the wavelength, $\lambda$, and the number of waves that pass a particular point in space per unit time is the frequency, $\nu$. The product of the wavelength and the frequency will be the speed of light, $c$, with which the wave travels.

$$\lambda \nu = c \qquad (8\text{-}1)$$

*Question 8.5*    Will a similar equation hold for waves produced on water surfaces? See Section 2 in "Handling the Phenomena" at the end of this chapter for directions on how to test this.

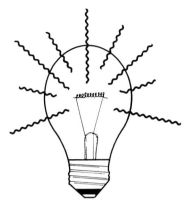

figure 8-7   The hot filament in an incandescent lamp, like any hot object, emits energy in the form of electromagnetic waves.

All electromagnetic waves travel at the same speed.[2] Therefore we can characterize a particular wave by either $\lambda$ or $\nu$. The various "types" of electromagnetic radiation and the $\lambda$'s associated with them are shown in Fig. 8-8. Actually they are all associated with the same physical phenomenon and we distinguish between visible, ultraviolet, and x-rays only because they manifest themselves to us in terms of the different effects that are produced by a particular range of wavelengths.

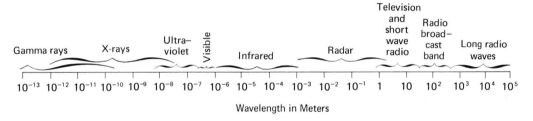

figure 8-8   The classifications of different wavelengths of electromagnetic radia-tion.

The "speed of light" $c$, is $3.0 \times 10^8$ meters/sec. What are typical values of $\nu$ for radio waves, infrared rays, visible light, and x-rays? The wavelengths for these regions are given in Fig. 8-8.

The problem attacked by Planck involved the theoretical explanation of the variation of intensity with wavelength, for radiation emitted by a hot

2. Actually this is strictly true only in a vacuum: in air there is a very slight change in speed for different wavelengths. In other substances the variation is somewhat larger, a property that enables us to use glass prisms to separate visible light into a spectrum of different wavelengths, each with its characteristic color.

object. Fig. 8-9 is a graphic presentation of an experimentally observed intensity $vs$ wavelength distribution for a radiating solid at different temperatures. The actual continuous spectrum produced by a radiating solid is shown in Fig. 8-12. Planck's predecessors assumed that it was the oscillatory motion of the charged particles in the heated object that produced the radiation. Their inability to explain the shape of the curves shown in Fig. 8-9 from what was known about the emission of electromagnetic waves by oscillating charge systems was puzzling indeed. All of these attempts were based on the assumption that the radiated energy was emitted continuously, with the intensity of the radiation being related to the amplitude of the oscillating particles. Planck's startling discovery was that a model, in which the energy was emitted in a discontinuous fashion in the form of minute bundles or quanta, could exactly reproduce the energy distribution curves. The magnitude of the quantum of energy was related to the frequency of oscillation by the formula:

$$\Delta E = h\nu \tag{8-2}$$

in which $h$ is a universal constant (now called Planck's constant) which has the value $6.625 \times 10^{-34}$ joule $\cdot$ second.

*Question 8.7*    What size quanta are associated with the emission of an x-ray from an oscillator with a frequency $10^{17}$ cycles per sec., and with the emission of a radio wave from an oscillator with a frequency of $10^6$ cycles per second?

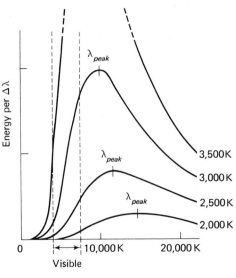

figure 8-9   Distribution of energy in the radiation emitted by a solid at each of four different temperatures. (From *Foundations of Modern Physical Science* by G. Holton and D. H. D. Roller. 1958. Addison-Wesley.)

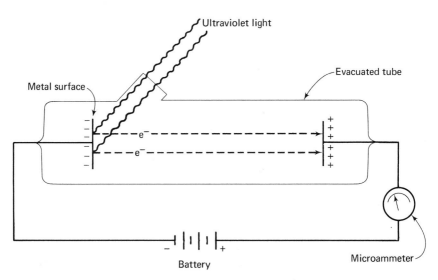

figure 8-10 The photoelectric effect. Ultraviolet light shining on the metal knocks off electrons, which travel through the vacuum tube to a positively charged plate. This produces a measurable electrical current.

The revolutionary notion that energy emission is quantized did not sit well with Planck's contemporaries. Indeed, even Planck doubted its validity and spent much time seeking an alternative. Another development, however, soon reinforced and extended the quantum idea. In 1888, the experimental physicist Heinrich Hertz had discovered a phenomenon now known as the photoelectric effect. If light of the appropriate frequency is allowed to strike the surface of certain metals enclosed in a vacuum tube, electrons are knocked off which can be observed in terms of an electric current between the negatively charged metal surface and second, positively charged metal plate, as illustrated in Fig. 8-10. According to the classical view of electromagnetic waves, the energy available should depend only on the intensity of the radiation and should be independent of its frequency. However, the energy of the electrons emitted in the photoelectric cell *is*, in fact, related to the frequency in the manner shown by Fig. 8-11. Below a certain threshold frequency, $\nu_0$, no electrons are emitted at all—*no matter how intense the light source*. At frequencies greater than the threshold frequency, an increase in intensity raises the number of electrons ejected, but has no effect on their energy! This startling result defied explanation until Albert Einstein published a paper in 1905 with the unassuming title, "On a Heuristic Point of View Concerning the Generation and Transformation of Light." In this paper Einstein proposed that not only is light emitted in quanta, as assumed by Planck, but that it somehow persists in the form of discrete bundles, with an energy equal to $h\nu$. He named these discrete bundles of energy *photons*. Thus when an electron in the metal surface

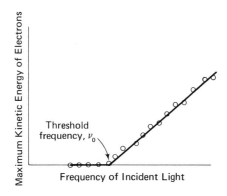

figure 8-11    The maximum kinetic energy of emitted photoelectrons graphed as a function of the incident light frequency.

interacts with electromagnetic radiation it can only absorb energy in a quantized fashion. For a given frequency, $\nu$, the electron receives energy equal to $h\nu$. Part of this energy is used to remove it from the potential energy well which results from the attraction between the negative electrons and the positive nuclei in the metal. If we symbolize this minimum energy needed to free an electron by $\phi$, then we obtain the equation proposed by Einstein, which agrees exactly with the observations illustrated in Fig. 8-11.

$$\text{Electron energy} = h\nu - \phi \qquad (8\text{-}3)$$

*Question 8.8*    If $\phi$ is $3.0 \times 10^{-19}$ joules for potassium, what is the threshold frequency for emitting electrons from that metal? What will be the speed of the electrons emitted by light with a frequency of $5.0 \times 10^{14}$ cycles/sec? (Note: You will have to recall the formula for kinetic energy and look up the mass of the electron.)

At frequencies smaller than $\nu_0$, where $h\nu_0 = \phi$, no electrons will be emitted because the quantum of energy carried by the photon is smaller than the energy needed to overcome the potential energy barrier, $\phi$. For radiation of higher frequency than $\nu_0$, the maximum energy of the emitted electron is given by equation 8-3. If, as proposed by Einstein, the number of photons of a particular frequency striking the metal surface is directly proportional to the intensity of the radiation, then the observed porportionality between the intensity and the number of electrons set free follows logically. Thus the puzzling experimental facts related to the photoelectric effect are neatly explained by these simple but revolutionary postulates concerning the nature of light energy.

Planck's and Einstein's work laid the foundation for the development of what has been termed the "new physics." The notion of energy quantization, as well as several other concepts completely foreign to the old classical mechanics of Newton et al., pervade this new quantum mechanics. It in no way invalidates the application of the previously derived relationships for macroscopic phenomena. Indeed, these relationships are derivable from the new, more general, theories. It should really not seem so surprising that the regularities observed for a bouncing baseball are not entirely applicable to the electron with approximately $10^{-30}$ of the mass.

*Question 8.9*  Is the mass ratio between an electron and a baseball roughly the same as that between the mass of a baseball and that of (a) The Empire State Building, or (b) The Earth, or (c) The Sun, or (d) The entire Milky Way Galaxy?

    In 1911 Bohr had these new controversial quantum ideas in his head when he attacked the problem of accounting for the stability of the nuclear atom. He also was very aware of other important regularities that linked the atom to the emission of electromagnetic radiation. Various workers in the 18th and 19th centuries had investigated the emission of light which resulted when certain metallic compounds were heated in a gas flame, and from gases through which electric sparks were passed. It was found that, unlike the continuous spectrum of light emitted by hot solids, energetically excited gas atoms emitted a complicated series of sharp spectral lines at definite frequencies as shown in Figure 8-12. The spectra were so specific that they soon came to be used to identify the presence of minute quantities of various elements. A frantic search for regularity in the gas spectra frustrated many researchers until a Swiss school teacher, Johann J. Balmer, announced in 1885 that he had found a regularity in the visible lines emitted by hydrogen atoms. Further work by Balmer, the Swedish scientist J. R. Rydberg, and others on light emitted from hydrogen in the ultraviolet and visible regions led to the general formula:

$$\frac{1}{\lambda} = R\left(\frac{1}{n_2^2} - \frac{1}{n_1^2}\right) \tag{8-4}$$

which predicts the exact wavelengths, $\lambda$, of all of the observed lines emitted by hydrogen atoms. The value of $R$ (known as the Rydberg constant) is $-1.097 \times 10^7$ m$^{-1}$ and $n_1$ and $n_2$ are any integers, 1, 2, 3 ..., with the restriction that $n_1$ must be smaller than $n_2$ so that $\lambda$ has a positive value.

*Question 8.10*  Calculate the wavelength and frequency of the emission line of atomic hydrogen characterized by $n_1 = 2$ and $n_2 = 4$. You can see this particular line as well as other spectra by following the instructions in Section 3 of "Handling the Phenomena" at the end of the chapter.

Bohr realized that any successful model of the hydrogen atom must not only explain the observations of Rutherford but must also elucidate this extraordinary regularity in the emission spectrum of the hydrogen atom. He noted that the emission of light at discrete wavelengths (or energies) suggested a quantization of the sort embodied in the theories of Planck and Einstein. He further observed that if the atom were restricted to changing energy by discrete amounts rather than continuously, the stability of a planetary type system could be justified because quantized changes in energy would preclude a continuous gradual loss of energy through radiation.

Bohr's 1911 model for the hydrogen atom consisted of an electron traveling in circular orbits around a central, relatively massive, nucleus with a single positive charge.

Bohr discovered that if he assumed that the angular momentum of the electron (given by the product of the mass $m$, velocity $v$, and radius, $r$) were restricted to discrete quantized multiples of a basic unit $h/2\pi$ ($h$ is Planck's constant), then the observed properties of the atom including the detailed emission spectrum were accounted for.

We can see how this works as follows: First the angular momentum is set equal to some integral multiple of $h/2\pi$.

$$mvr = \frac{nh}{2\pi} \tag{8-5}$$

Next we equate the centripetal force, given by $mv^2/r$, with the inward electrostatic attraction. This attraction is equal to $(-kq_1q_2)/r^2$ where $q_1$ and $q_2$ are the charges on the nucleus and electron, $r$ is the distance between the nucleus and the electron and $k = 8.988 \times 10^9 \text{ N} \cdot m^2/\text{coulomb}^2$. For our purposes here, we will consider the coulomb as being simply a unit quantity of electricity.

$$\frac{mv^2}{r} = \frac{-kq_1q_2}{r^2} \tag{8-6}$$

If we solve Equation 8-5 for $v$:

$$mvr = \frac{nh}{2\pi}$$

$$2\pi mvr = nh$$

$$v = \frac{nh}{2\pi mr}$$

Substituting this result in Equation 8-6, we obtain:

$$\frac{m\left(\dfrac{nh}{2\pi mr}\right)^2}{r} = \frac{-kq_1q_2}{r^2}$$

Continuous spectrum

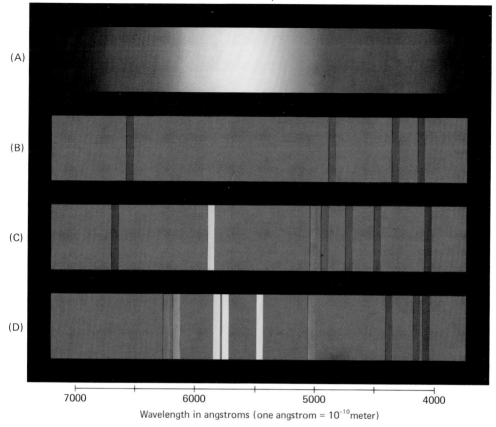

(A)

(B)

(C)

(D)

Wavelength in angstroms (one angstrom = $10^{-10}$ meter)

figure 8-12 The continuous visible emission spectrum of a hot tungsten ribbon (A); and the visible line spectra of (B) hydrogen atoms, (C) helium atoms, and (D) mercury atoms. [(A) and (B); From *Chemistry: An Experimental Science*, G.C. Pimentel, ed. Copyright © 1963. W. H. Freeman and Co.; (C) and (D); *From Chemistry: Man and Matter* by E. R. Hardwick and C. Knobler. 1970. Ginn.]

$$\frac{m\left(\dfrac{n^2h^2}{4\pi^2m^2r^2}\right)}{r} = \frac{-kq_1q_2}{r^2}$$

$$\left(\frac{m}{r}\right)\left(\frac{n^2h^2}{4\pi^2m^2r^2}\right) = \frac{-kq_1q_2}{r^2}$$

$$\left(\frac{n^2h^2}{4\pi^2mr^3}\right) = \frac{-kq_1q_2}{r^2} \tag{8-7}$$

Solving for $r$ gives the result:

$$\frac{n^2h^2}{4\pi^2mr^3} = \frac{-kq_1q_2}{r^2}$$

$$r^2\left(\frac{n^2h^2}{4\pi^2mr^3}\right) = -kq_1q_2$$

$$\frac{n^2h^2}{4\pi^2mr} = -kq_1q_2$$

$$\frac{n^2h^2}{4\pi^2m} = -kq_1q_2r$$

$$\frac{\dfrac{n^2h^2}{4\pi^2m}}{-kq_1q_2} = r$$

$$\frac{n^2h^2}{-kq_1q_24\pi^2m} = r \tag{8-8}$$

If we define a new constant $a_0 = -(h^2/kq_1q_24\pi^2m)$ we can rewrite Equation 8-8 in the form:

$$r = n^2a_0 \tag{8-9}$$

This means that the radius of the stable circular orbits for the electron in the Bohr model for the hydrogen atom is restricted to the discrete values $a_0$, $4a_0, 9a_0, 16a_0, \ldots$, corresponding to values of the $n$ "quantum number" equal to 1, 2, 3, 4, . . . as shown in Fig. 8-13.

*Question 8.11*   Show that the smallest allowed radius for the electron in the hydrogen atom, $a_0$, is $5.29 \times 10^{-11}$ m. (Note: the electric charge on the electron is $-1.602 \times 10^{-19}$ coulombs).

To obtain an expression for the energy of the hydrogen atom we first add the expressions for the kinetic energy, $\frac{1}{2}mv^2$, and potential energy, $kq_1q_2/r$ to give

$$E = \tfrac{1}{2}mv^2 + k\frac{q_1q_2}{r} \tag{8-10}$$

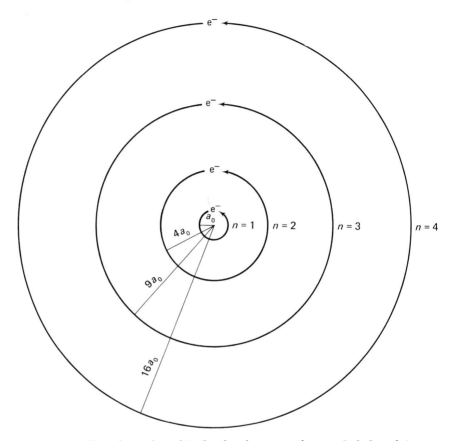

figure 8-13    Allowed circular orbits for the electron is the $n = 1, 2, 3,$ and 4 states of the Bohr model for the hydrogen atom.

From Equation 8-6

$$\frac{mv^2}{r} = \frac{-kq_1q_2}{r^2}$$

we see that

$$mv^2 = \frac{r(-kq_1q_2)}{r^2}$$

so that:

$$mv^2 = \frac{-kq_1q_2}{r}$$

Substituting this last equality in Equation 8-10 we obtain

$$E = \tfrac{1}{2}mv^2 + k\frac{q_1q_2}{r}$$

$$E = \tfrac{1}{2}\left(\frac{-kq_1q_2}{r}\right) + \left(\frac{kq_1q_2}{r}\right)$$

$$E = \tfrac{1}{2}\left(\frac{kq_1q_2}{r}\right)$$

$$E = \frac{kq_1q_2}{2r} \tag{8-11}$$

If we next substitute the expression given by Equation 8-9 for $r$, we find that

$$E = k\frac{q_1q_2}{2(n^2a_0)} \tag{8-12}$$

Bohr thus obtained a model for the hydrogen atom having an orbiting electron restricted to a set of states, $n$, each corresponding to a definite orbital radius. These radii are given by Equation 8-9, and have the quantized energy specified by Equation 8-12. When an atom emits electromagnetic energy, the electron must be "falling" from some high energy state, $n_2$, to a lower energy state, $n_1$, (having a smaller radius). We can use Equation 8-12 to find the corresponding energy difference, $E_2 - E_1$, between the two states.

$$E_2 - E_1 = k\frac{q_1q_2}{2a_0}\left[\frac{1}{n_2^2} - \frac{1}{n_1^2}\right] \tag{8-13}$$

If we further assume that the energy is radiated in the form of a photon of energy $h\nu$ and recall that $\lambda\nu = c$ we can write an expression for the wavelength of the light emitted:

$$\frac{hc}{\lambda} = k\frac{q_1q_2}{2a_0}\left[\frac{1}{n_2^2} - \frac{1}{n_1^2}\right] \text{ or } \frac{1}{\lambda} = \frac{kq_1q_2}{hc2a_0}\left[\frac{1}{n_2^2} - \frac{1}{n_1^2}\right] \tag{8-14}$$

This is recognizable as having the same form as the experimentally derived Equation 8-4! If we plug in the values of $k$, $q_1$, $q_2$, $h$, $c$, and $a_0$ we find that $kq_1q_2/hc2a_0$ is indeed equal in value to the Rydberg constant!

Question 8.12   Verify that $kq_1q_2/hc2a_0 = -1.097 \times 10^7$ m$^{-1}$. (Note: the minus sign results from the fact that the potential energy is defined as being equal to zero when $r$ becomes infinitely large. Therefore, the lowering of the electron's potential energy, due to nucleus-electron attraction, results in negative values for the energies of the states of the atom.)

This remarkable achievement, which accounted for the stability and spectrum of the hydrogen atom, was amplified by the successful prediction of other properties of this simplest of atoms. Unfortunately, attempts by Bohr and others to apply his model, even in considerably modified form, to other atoms met with only limited success.

Our present understanding of atomic structure and other phenomena involving submicroscopic particles depends on the work of many other founders of the new physics such as Werner Heisenberg, Erwin Schrödinger, Wolfgang Pauli, Louis de Broglie, Max Born and Paul Dirac. To the notions of quantization of energy and momentum many other concepts have been added, which appear strange to anyone who has confined their attention solely to the physics of macroscopic systems. For example, "particles" as small as the electron can no longer be thought of as simultaneously having a well defined position and a specific momentum. In place of circular orbits for the electrons we find that we can only attribute to each state a probability distribution which tells us the likelihood of finding an electron at a certain position in space, but says nothing about a specific path. These probability distributions have some of the characteristics of waves. The mathematical equations, although perfectly adequate for calculating properties of interest, defy translation into simple visual models that have analogies to phenomena with which we are all familiar.

In place of the single quantum number, $n$, introduced by Bohr, we find that each allowed state of an electron in an atom requires four such numbers for its description. The so-called principal quantum number, designated by the symbol $n$, and taking on integral values, 1, 2, 3, 4 . . . , determines how spread out in space the electron distribution is and also is the major determinant of the energy. A second quantum number, $l$, which can have integral values beginning with zero, but not exceeding $(n - 1)$, determines the angular momentum, angular shape of the spatial distribution, and (for all atoms other than hydrogen) also contributes to determining the electron's energy. A third quantum number, usually designated as $m_l$, describes how a given state will be affected by external magnetic fields, and can have values ranging from $-l$ to $+l$. The significance of the final quantum number, $m_s$, can be visualized if we picture the electron as a little top which can spin on its axis in either a clockwise ($m_s = +\frac{1}{2}$) or a counterclockwise ($m_s = -\frac{1}{2}$) fashion. A pictorial representation of spatial distributions for electrons in a few states is given in Fig. 8-14. In the following section we shall see how this scheme is the basis for understanding the physical and chemical properties of the elements.

## 8.4   atomic periodicity

Since the very essence of scientific activity is a search for order in the universe, it is not surprising that the discovery of the chemical elements resulted in an intensive search for regularities in their properties. The elements seemed to be in families—metals, gases, etc. The atomic theory provided a basis for such studies, in terms of the one to one relationship between the set of elements and the set of atoms. Each element corresponds to a given atom with a given

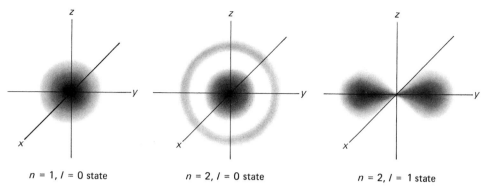

figure 8-14   Spatial probability distributions for electrons in the first few quantum states. (From *Chemistry—An Experimental Science*, G. C. Pimentel, ed. Copyright 1963 by W. H. Freeman and Co.)

set of properties. It was periodicity in the physical and chemical properties of the elements and certain of their compounds, as well as the relative masses of the individual atoms that served as the basis for the initial successes in the quest for order among the elements. The Russian chemist, Dmitri Mendeleev, and the German, Julius Meyer, independently published papers in the 1870's which clearly demonstrated that the elements could be arranged in tables, based upon increasing atomic mass, in such a way as to display marked periodicity in certain of their properties. The most remarkable achievement of these two scientists was their ability to predict successfully the existence of undiscovered elements on the basis of obvious gaps which appeared in their tables. Later work by the English physicist, Henry G-J. Moseley showed that the proper variable upon which to base the periodic table was not the atomic mass, but rather an integer derived from x-ray spectra of the elements which he called the *atomic number*. This integer is now known to be equal to the number of positively charged protons in the atomic nucleus which is the same as the number of negative electrons in the neutral atom. Thus hydrogen with one proton and one electron has atomic number one, whereas carbon with six of each is assigned number six, etc. (The reason that Mendeleev's and Meyer's arrangements were so successful is that with only three exceptions the elements, when listed by increasing atomic mass, have also been ordered by increasing atomic number.)

A modern periodic table of the elements is shown on page 212. The arrangement is such that elements in vertical columns resemble each other in chemical and physical properties. In general the elements tend to become less metallic in going across a row from left to right and in going up a column.

As a first example of the periodic behavior of the elements, we find as shown in Fig. 8-15 that the elements with atomic numbers 2, 10, 18, 36, 54, and 86 are gases which are relatively inert, chemically. (They were once thought to be completely unreactive but, within recent years, some of them have been found to undergo a limited number of reactions with highly reactive elements.)

| | 2<br>He |
|---|---|
| | 10<br>Ne |
| | 18<br>Ar |
| | 36<br>Kr |
| Xenon | 54<br>Xe |
| Radon | 86<br>Rn |

*(Helium, Neon, Argon, Krypton, Xenon, Radon listed at left)*

figure 8-15  The noble gas elements, helium, neon, argon, krypton, xenon, and radon. (From *Chemistry: An Experimental Science*, G. C. Pimentel, ed Copyright © 1963. W. H. Freeman and Co.)

In the periodic table (such as that shown in Fig. 8-16) each of these elements is followed by a very soft shiny reactive metallic element called an *alkali metal*. With the exception of the element with atomic number 2 (helium), the other relatively inert gases are preceded in the table by elements known as halogens. The halogen elements are all reactive and nonmetallic.

Those who probed the structure of the atom in the first quarter of this century had, as one of their most challenging tasks, the job of explaining chemical periodicity. Most scientists today believe that the wave mechanical theory which incorporates probability distributions, wave properties, and quantization in its equations for the description of physical systems provides

| 1<br>H | | | | | | | | | | | | | | | | | 2<br>He |
|---|---|---|---|---|---|---|---|---|---|---|---|---|---|---|---|---|---|
| 3<br>Li | 4<br>Be | | | | | | | | | | | 5<br>B | 6<br>C | 7<br>N | 8<br>O | 9<br>F | 10<br>Ne |
| 11<br>Na | 12<br>Mg | | | | | | | | | | | 13<br>Al | 14<br>Si | 15<br>P | 16<br>S | 17<br>Cl | 18<br>Ar |
| 19<br>K | 20<br>Ca | 21<br>Sc | 22<br>Ti | 23<br>V | 24<br>Cr | 25<br>Mn | 26<br>Fe | 27<br>Co | 28<br>Ni | 29<br>Cu | 30<br>Zn | 31<br>Ga | 32<br>Ge | 33<br>As | 34<br>Se | 35<br>Br | 36<br>Kr |
| 37<br>Rb | 38<br>Sr | 39<br>Y | 40<br>Zr | 41<br>Nb | 42<br>Mo | 43<br>Tc | 44<br>Ru | 45<br>Rh | 46<br>Pd | 47<br>Ag | 48<br>Cd | 49<br>In | 50<br>Sn | 51<br>Sb | 52<br>Te | 53<br>I | 54<br>Xe |
| 55<br>Cs | 56<br>Ba | 57<br>*La | 72<br>Hf | 73<br>Ta | 74<br>W | 75<br>Re | 76<br>Os | 77<br>Ir | 78<br>Pt | 79<br>Au | 80<br>Hg | 81<br>Tl | 82<br>Pb | 83<br>Bi | 84<br>Po | 85<br>At | 86<br>Rn |
| 87<br>Fr | 88<br>Ra | 89<br>†Ac | | | | | | | | | | | | | | | |

| * | 58<br>Ce | 59<br>Pr | 60<br>Nd | 61<br>Pm | 62<br>Sm | 63<br>Eu | 64<br>Gd | 65<br>Tb | 66<br>Dy | 67<br>Ho | 68<br>Er | 69<br>Tm | 70<br>Yb | 71<br>Lu |
|---|---|---|---|---|---|---|---|---|---|---|---|---|---|---|
| † | 90<br>Th | 91<br>Pa | 92<br>U | 93<br>Np | 94<br>Pu | 95<br>Am | 96<br>Cm | 97<br>Bk | 98<br>Cf | 99<br>Es | 100<br>Fm | 101<br>Md | 102<br>No | 103<br>Lw |

figure 8-16  Periodic table of the elements.

the basis for calculating all of the observed properties of atoms. Unfortunately, the calculations for all but the simplest cases are extremely complex and lengthy. The nature of the problem is such that as the number of electrons increases, the complexity of the problem of determining a particular atomic property increases exponentially. For atoms or molecules with more than four or five electrons, even the largest fastest computers are not up to the task. From the success achieved with the simple cases, and the approximate results obtainable for the other atoms, it seems that the theory is valid and adequate, and we can explain qualitatively why the properties of atoms display an orderly repetitiveness.

In Section 8.3 the four quantum numbers which specify the possible energy states of electrons in atoms were described. The lowest energy states are those for which $n = 1$. For this value of $n$ we find that $l$ and $m_l$ can only be zero, while $m_s$ can be either $+\frac{1}{2}$ or $-\frac{1}{2}$. Thus, since the energy depends only on the value of $n$ and $l$, these two states have the same energy. They can be designated by the quantum numbers $n = 1$, $l = 0$, $m_l = 0$, $m_s = 1/2$ and $n = 1$, $l = 0$, $m_l = 0$, $m_s = -1/2$. One might expect that the lowest energy state of any atom is one in which all of the electrons, simultaneously, are in one of these two lowest energy states. However, this is not the case! An additional restriction, first recognized by Wolfgang Pauli, is an important part of the new physics. As applied to atoms, this restriction (known as the *Pauli exclusion principle*) states that no two electrons can simultaneously be in any one quantum state. Another way of stating it is that for any two electrons in a given atom, at least one of their four quantum numbers must be different.

Let us examine the consequences of the above discussion with respect to the lowest energy states of the first few atoms in the periodic table. Hydrogen will have its single electron in one of the two $n = 1$ states. The next element, helium, has one electron in each of the two $n = 1$ states. Lithium, the third element, requires separate states for each of its three electrons and since there are only two $n = 1$ states, one of its electrons is forced to occupy a higher energy state. For $n = 2$ we find that the following states are available:

| $n$ | $l$ | $m_l$ | $m_s$ |
|---|---|---|---|
| 2 | 0 | 0 | $\frac{1}{2}$ |
| 2 | 0 | 0 | $-\frac{1}{2}$ |
| 2 | 1 | $-1$ | $\frac{1}{2}$ |
| 2 | 1 | $-1$ | $-\frac{1}{2}$ |
| 2 | 1 | 0 | $\frac{1}{2}$ |
| 2 | 1 | 0 | $-\frac{1}{2}$ |
| 2 | 1 | 1 | $\frac{1}{2}$ |
| 2 | 1 | 1 | $-\frac{1}{2}$ |

*Question 8.13*    Verify that these are the only possible states with $n = 2$ in terms of the restrictions on the quantum numbers given in Section 8.3.

The two states with $n = 2$, $l = 0$ will be at a somewhat lower energy than the six states with $n = 2$, $l = 1$ due to the dependence of the energy on the value of $l$ as well as on $n$. The eight electrons that are successively added in proceeding from lithium through neon are accommodated by these eight quantum states. The remainder of the elements are "constructed" in this way by adding each additional electron to the lowest-energy unoccupied quantum state. A schematic diagram showing the ordering of all of the states up to and including those with $n = 3$ is given in Fig. 8-17. (Note that the two $n = 4$, $l = 0$ states occur at lower energies than the $n = 3$, $l = 2$ states.)

*Question 8.14*    What element will have been reached when all of the $n = 3$ states are completely filled? (Hint: note that each $l = 0$ state can accommodate two electrons [one with $m_s = \frac{1}{2}$ and one with $m_s = -\frac{1}{2}$]. An $l = 1$ state can accommodate six electrons [three possible values of $m_l$, each with two values for $m_s$]. Similarly an $l = 2$ state can contain ten electrons.)

To see how this numbers game leads to a qualitative explanation of periodic behavior we will consider one of the most important of atomic properties, the *ionization energy*. This quantity is defined as the minimum energy needed to completely remove the outermost electron from a given atom. (The resulting species has a $+1$ electronic charge and is called an *ion*.) The ionization energy is a measure of how tightly the electron is bound to the atom. Comparing helium to hydrogen we find that the nuclear charge has doubled. Each electron in helium will "feel" the attraction from a $+2$ nucleus. Thus, the ionization energy of helium is much higher than that of hydrogen. Proceeding to lithium, we find that the nuclear charge is up to $+3$ but the highest energy

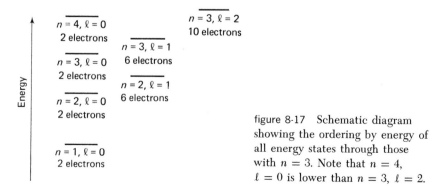

figure 8-17    Schematic diagram showing the ordering by energy of all energy states through those with $n = 3$. Note that $n = 4$, $l = 0$ is lower than $n = 3$, $l = 2$.

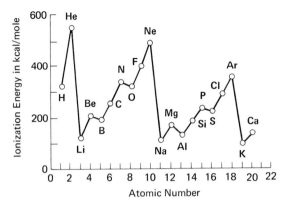

figure 8-18    The periodicity of the ionization energy
for the first 20 elements. (From *Chemistry—An Experimental Science*, G. C. Pimentel, ed. Copyright
© 1963. W. H. Freeman and Co.)

electron is now in an $n = 2$ state much further out, on the average, from
the nucleus than the two $n = 1$ electrons. The lithium nucleus, with three
units of positive charge, is "seen" by the high energy electron as being
"shielded" by the two lower energy electrons. Since each of these carries
one unit of negative charge, the result is that the high energy electron "sees"
and "effective" core charge of about $+1$. Thus, the ionization energy of
lithium is much *lower* than that of helium. Since the $n = 2$ electron of lithium
is also considerably further from the nucleus than the hydrogen electron, its
ionization energy is also somewhat lower than that of hydrogen. From lithium
to neon the nuclear charge successively increases by $+1$ while filling $n = 2$
states. Thus, the ionization energy increases regularly. The element following
neon in the periodic table is sodium, for which we find that the highest energy
electron is forced to occupy an $n = 3$ state. It is still farther away from the
nucleus and shielded from its attraction by ten other electrons. Just as in the
case of lithium a marked drop in ionization energy can be predicted (and
found). This process repeats itself throughout the table. Each time one of the
relatively inert gases is reached a high effective core charge has been built
up, resulting in a large ionization energy. The following element (an alkali
metal) always has an electron in the next $n$ state with a resulting dramatic
drop in ionization energy. This periodicity of the ionization energy is illustrated in Fig. 8-18.

The specific interactions which occur between atoms, which is called
chemical reactivity, involves changes in the probability distribution of the
electrons. As we shall see, these rearrangements might require the complete
or partial removal of the highest energy electrons of a reacting atom. The
higher the ionization energy, the more difficult it is to remove an electron.
The relative inertness of elements 2, 10, 18, 36, 54, and 86 is, in part, due

to their high ionization energies and the reactivity and metallic properties of elements 3, 11, 19, 37, 55, and 87 is dependent on the relative ease with which their outermost electrons may be removed.

In the next chapter we shall examine further examples of periodic behavior of the chemical elements and the explanations in terms of the wave mechanical model of the atom.

## summary

In this chapter we examined the atomicity of matter—a model which had its crude beginnings as early as the fifth century B.C. We saw how this controversial model was given a firm quantitative foundation in the atomic theory of John Dalton. His postulates were successful in explaining a considerable body of experimental data on definite proportions and in formulating the law of multiple proportions.

The discovery that atoms are composed of still smaller building blocks led to a search for a model to explain the structure of the atom. Careful experimentation revealed that atoms consist of a tiny positively charged nucleus, containing most of the mass of the atom, surrounded by negatively charged electrons.

The instability of the classical planetary model of the atom contributed to the birth of the "new physics" which attributes wave-like behavior to subatomic particles, and particle-like behavior to light waves. In addition to providing a successful model for the atom this new quantum theory explains such other related phenomena as the photoelectric effect and the line spectra of the elements.

From intensive study of regularities in the spacing of emission lines in the spectrum of hydrogen atoms, and from much other data, scientists were able to derive our modern model of the atom in which the quantized electronic energy states are described by four quantum numbers.

The observed periodicity in the properties of elements when listed according to increasing atomic number (number of protons, positively charged particles, in the nucleus) was shown to follow from the "exclusion principle" which says that no two electrons in an atom can simultaneously occupy the same quantum state as specified by the values of the four quantum numbers.

*Answers to*    8.1    If the fraction of the 79.9% copper-containing oxide is $x$, then $(1 - x)$ will be
*Questions*    the fraction of the 88.8% copper-containing oxide. Thus:

$$79.9x + 88.8(1 - x) = 83.1$$

$$8.9x = 5.7$$

$$x = .64 = \text{fraction of 79.9\% compound}$$

$$(1 - x) = .36 = \text{fraction of 88.8\% compound}$$

8.2   According to postulate (4) a given compound will have as its basic unit a molecule which consists of a fixed number of atoms of each of the elements of which it is composed. Postulate (3) states that the atoms of each of the elements will have a definite characteristic mass. Therefore, the relative porportions of each element in the compound will be fixed by the product of the number of atoms of that element in the compound and the unique atomic mass of that element. For instance, if a molecule of water always contains two atoms of hydrogen and one atom of oxygen, and if an oxygen atom has 16 times the weight of a hydrogen atom, then water will always contain the definite proportion by weight of 8:1 for oxygen and hydrogen.

8.3   Suppose you shot a small positively charged pellet (alpha particle) at a pellet that was about fifty times as massive and carried a positive charge 39.5 times larger (gold atom nucleus). If you shot randomly the chance of hitting the large pellet wouldn't be very great. In the event that you did score a hit and small pellet would undergo a large deflection.

   If you ground up the large pellet into a fine charged dust and distributed this dust over a large volume (raisin pudding model) and fired your small pellet into it, there would be no way in which a large deflection in path would be produced.

8.4   If the orbiting particle radiates energy its total energy will have to decrease in order to maintain the overall conservation of mass-energy. As indicated in Chapter 6, a decrease in total energy of an orbiting particle requires a decrease in its orbital radius. Thus, as the electron continues to radiate energy, it will spiral into the nucleus.

8.5   The equation holds, but the "$c$" that you measure will be much smaller than the speed of electromagnetic waves.

8.6   For a typical radio wave, $\lambda = 10^2$ meters. Therefore:

$$\nu = \frac{3.0 \times 10^8}{10^2} = 3 \times 10^6/\text{sec}$$

For a typical infrared ray, $\lambda = 10^{-5}$ meters. Therefore:

$$\nu = \frac{3.0 \times 10^8}{10^{-5}} = 3 \times 10^{13}/\text{sec}$$

For a typical visible light wave, $\lambda = 5 \times 10^{-7}$ meters. Therefore:

$$\nu = \frac{3 \times 10^8}{5 \times 10^{-7}} = 6 \times 10^{14}/\text{sec}$$

For a typical x-ray, $\lambda = 10^{-9}$ meters. Therefore:

$$\nu = \frac{3 \times 10^8}{10^{-9}} = 3 \times 10^{17}/\text{sec}$$

8.7   For the x-ray,

$$\Delta E = h\nu = (6.625 \times 10^{-34} \text{ joule} \cdot \text{sec})(10^{17}/\text{sec}) = 6.625 \times 10^{-17} \text{ joule}$$

For the radio wave,

$$\Delta E = h\nu = (6.625 \times 10^{-34} \text{ joule} \cdot \text{sec})(10^6/\text{sec}) = 6.625 \times 10^{-28} \text{ joule}$$

Both of these are small amounts of energy on a human scale, but note that the x-ray quantum has $10^{11}$ (100 billion) times the energy of the radio quantum.

8.8   Threshold frequency will emit electrons with zero energy:

$$0 = h\nu_0 - \phi \text{ or } \phi = h\nu_0$$

$$3.0 \times 10^{-19} \text{ joules} = (6.625 \times 10^{-34} \text{ joule} \cdot \text{sec})(\nu_0)$$

$$\nu_0 = 4.5 \times 10^{14}/\text{sec}$$

For light of frequency, $5.0 \times 10^{14}$ (blue-green in color)

$$\text{Electron energy} = (6.625 \times 10^{-34})(5.0 \times 10^{14}) - (3.0 \times 10^{-19})$$

$$= 3.0 \times 10^{-20} \text{ joule}$$

The formula for kinetic energy is $E = \frac{1}{2}mv^2$ and $m$ for an electron is $9.11 \times 10^{-31}$ kg. Therefore:

$$3 \times 10^{-20} = \frac{1}{2}(9.11 \times 10^{31})v^2$$

$$v^2 = 6.6 \times 10^{10}$$

$$v = 2.6 \times 10^5 \text{ m/sec or } 260 \text{ km/sec}$$

8.9   The ratio of the mass of a baseball ($1.5 \times 10^{-1}$ kg) to that of an electron ($9 \times 10^{-31}$ kg) is approximately $10^{30}$ to 1. Thus we are looking for an object that has a mass on the order of $10^{30}$ kg. The correct answer is the sun, which has a mass of $2 \times 10^{30}$ kg. (The mass of the earth is approximately $6 \times 10^{24}$ kg—clearly too small!)

8.10

$$\frac{1}{\lambda} = (1.097 \times 10^7)\left(\frac{1}{2^2} - \frac{1}{4^2}\right)$$

$$= 2.057 \times 10^6/\text{m}$$

$$\lambda = 4.862 \times 10^{-7} \text{ m}$$

$$\nu = \frac{c}{\lambda} = \frac{2.998 \times 10^8}{4.862 \times 10^{-7}} = 6.166 \times 10^{14}/\text{sec}$$

8.11

$$a_0 = -\frac{h^2}{kq_1q_24\pi^2m}$$

substituting the values for $h$, $k$, $q_1q_2$, $\pi$ and $m$ we have (note: $q_1 = -q_2$):

$$a_0 = -\frac{(6.625 \times 10^{-34})^2}{(8.988 \times 10^9)(-1.602 \times 10^{-19})(1.602 \times 10^{-19})(4)(3.142)^2(9.109 \times 10^{-31})}$$

$$= 5.29 \times 10^{-11} \text{ m}$$

Since this would be the radius of the electron orbit in an ordinary room temperature hydrogen atom, the diameter would be about $10 \times 10^{-11}$ m, or $10^{-10}$ m. How does the result of this calculation compare with the experimental estimate of atomic size made in Chapter 7?

8.12

$$\frac{kq_1q_2}{hc2a_0} = \frac{(8.988 \times 10^9)(-1.602 \times 10^{-19})(1.602 \times 10^{-19})}{(6.625 \times 10^{-34})(2.998 \times 10^{10})(2)(5.29 \times 10^{-11})} = -1.097 \times 10^7/\text{m}$$

8.13   For $n = 2$, $l$ can only have values of 0 and 1 (it can not exceed $(n - 1)$) For $l = 0$ $m_l$ can only be 0. For $l = 1$, $m_l$ can be $-1$, 0, or $+1$. For each possible value of $l$ and $m_l$, $m_s$ can be $+\frac{1}{2}$ or $-\frac{1}{2}$. Thus we get only those eight states listed.

8.14   If we count the number of states in Fig. 8-17 making use of the fact that each $l = 0$, $l = 1$, or $l = 2$ state can accommodate two, six, or ten electrons respectively, we find that they can accommodate 30 electrons. Element number 30 is zinc.

handling the phenomena

1.  *Definite Proportions*
**CAUTION!** In this experiment you will be working with a very corrosive substance, hydrochloric acid. The following safety precautions should be strictly observed.

1   Wear safety goggles or glasses.
2   Take care not to splash the acid on your skin or clothing.
3   If you do have an accident, wash away the spilled acid and flush any exposed skin areas with running water for several minutes.
4   The evaporation of the acid will produce caustic HCl fumes. This step should preferably be performed in a fume hood. If a hood is not available, a well-ventilated area near an open window can be used. If the evolution of HCl gas becomes sufficient to produce noticeable eye or lung irritation, remove the solution from the source of heat immediately and evacuate the affected area.

The law of definite proportions can be demonstrated by observing that the ratio of weights of zinc and chlorine that combine to form the compound, zinc chloride, is fixed. This information will be extracted from results obtained when a measured amount of zinc reacts with excess hydrochloric acid solution. The products will be hydrogen gas and a solution containing the zinc chloride plus the excess hydrochloric acid. The hydrogen that is produced will bubble out of the solution and the volatile hydrochloric acid will be driven off by heating. The nonvolatile solid zinc chloride will be left behind after the liquid evaporates.

Carefully weigh a clean dry evaporating dish on a triple beam balance. Record the weight. Next add a few particles of granular zinc metal. Again weigh the dish and find the weight of the zinc by subtracting the first weight from the second. Record this weight of the zinc. (The sample of zinc should weigh between one and five grams—if not, add or subtract some zinc and weigh it again.)

Next, add 15 ml (one ml is equal to one $\text{cm}^3$) of dilute hydrochloric acid to the zinc. THIS SHOULD BE DONE SLOWLY, WHILE STIRRING. The reaction that you observe results in the formation of hydrogen gas (note the bubbles) and zinc chloride solution. If there is any zinc left after the reaction stops, add an additional 5 ml of hydrochloric acid. Repeat until you have a clear solution.

Now form a little double boiler by placing the evaporating dish on top of a small beaker containing some water. Bring the water in the beaker to a boil by heating it on a hot plate or a Bunsen burner flame and continue to heat until the liquid in the evaporating dish has almost disappeared. Remove the evaporating dish from the water beaker, place it on a wire gauze and heat it until a dry powder is obtained. (During this direct heating care must be taken to avoid spattering. This can be done by using a low setting on the hot plate or by moving the flame to avoid excess local heating.)

Let the evaporating dish cool completely, and weigh its contents. All the zinc originally present on the dish has combined chemically with chlorine, forming the white zinc chloride powder. If you subtract the combined weight of the dish and the zinc from the final total weight you will be able to determine how much chlorine combined with your sample of zinc.

Calculate the ratio of the masses of chlorine to zinc in your sample of zinc chloride. Compare your results to those obtained by the other students. If you find a constant ratio (within experimental error) independent of the amount of zinc used or hydrochloric acid added, you have evidence of definite proportions for this compound of zinc and chlorine.

## 2.   *Waves on Water*

Fill a large plastic basin with a few centimeters of water. Place the basin directly beneath a light. After the water has become still, create waves by oscillating the water with your finger. Using a watch, determine the frequency of the waves produced by counting the number of finger oscillations per second. Practice until you can wiggle your finger at a steady frequency of about 2 oscillations per second.

You should be able to see the shadows of the waves produced on the bottom of the basin. Measure the wavelength (distance between wavecrest shadows) to the nearest half-centimeter. (This will require a second person and some practice.) Next, using a stopwatch, measure the speed with which the waves travel by determining how long a wave shadow takes to travel a given distance. (The distance should be at least 30 cm.)

Check to see if the relationship $\lambda \nu = v$ holds for water waves. (With this semi-quantitative method do not expect agreement to better than a factor of 2. If a laboratory ripple tank is available, you can get better precision.)

## 3.   *Emission Spectra*

Light can be separated into its component frequencies by passing it through either a triangular prism or a transmission grating consisting of closely spaced grooves. Such transmission gratings are available in the form of inexpensive $2'' \times 2''$ slides obtainable from the Edmund Scientific Company, Barrington, New Jersey.

Cut a slit about 0.5 mm wide and 2 cm long in a $5'' \times 8''$ index card. Hold a grating in front of one eye and look through it at an incandescent light bulb. Now place the card with the slit between the light source and the grating. You will see the continuous spectrum of light emitted by the bulb, spread out on both sides of the slit. (The card with the slit should be about 20–25 cm from your eye. If the source of light is a line or a point instead of a diffuse surface, you can do without the slit. Just hold the grating close to your eye and look at the source directly.)

Now look at atomic spectral light sources (obtainable from various laboratory suppliers) with the grating and slit. These light sources are made by passing an electric discharge through a low pressure gas of the element in question. Look at the spectra of hydrogen, sodium, mercury, or whatever other sources are available. Sketch the bright line pattern observed in each case. Now look at a fluorescent light. You should be able to observe both a continuous emission produced by the glowing solid coating of the bulb and some bright lines superimposed on it. The bright lines are due to the emission from an element present in the gas within the fluorescent tube. Compare the bright line spectrum to those of the elements you sketched. What element is producing the emission lines in fluorescent lights?

## problems

1   Water and hydrogen peroxide are both oxides of hydrogen containing 11.1 percent and 5.88 percent hydrogen, respectively, by weight. What is the weight percentage of hydrogen in a mixture of 70 percent water and 30 percent hydrogen peroxide?

2   Molecules of the compound nitrous oxide (laughing gas) contain two atoms of nitrogen for each atom of oxygen. In the compound nitrogen dioxide the ratio is reversed. (Two oxygen atoms per nitrogen atom). What will be the ratio of the masses of oxygen that will combine with a specified mass of nitrogen to form these two compounds?

3   In the next chapter you will learn that contrary to Dalton's atomic theory, all atoms of a given element do not necessarily have the same mass. It *is* true however, that the *average* mass of the atoms in a naturally occurring sample of a particular element is constant. Would accurate data on definite or multiple proportions reveal this distinction?

4   Why must the sum of the negative charges of the electrons of a neutral atom be equal to the positive charge of its nucleus?

5   Visible light of frequency $\nu = 4.7 \times 10^{14}$ sec$^{-1}$ appears red to human observers whereas light of frequency $\nu = 6.8 \times 10^{14}$ sec$^{-1}$ is seen as blue. What are the corresponding wavelengths for these two frequencies?

6   A certain metal is found to emit electrons with maximum kinetic energy of $4.5 \times 10^{-20}$ joules when light with a frequency of $5.5 \times 10^{14}$ sec$^{-1}$ strikes its surface in a vacuum. What is the value of $\phi$ (the minimum energy needed to free an electron) for this metal?

7   What distinguished Bohr's model of the hydrogen atom from the classic planetary model? How does our present model of the atom differ from the Bohr atom?

8   Which of the following quantum number designations of electron states are actually "allowed"?

| $n$ | $\ell$ | $m_\ell$ | $m_s$ |
|-----|--------|----------|-------|
| 1 | 1 | 0 | $\frac{1}{2}$ |
| 3 | 1 | $-1$ | $\frac{1}{2}$ |
| 0 | 1 | 1 | $\frac{1}{2}$ |
| 9 | 0 | 0 | $-\frac{1}{2}$ |
| 4 | 2 | 3 | $\frac{1}{2}$ |
| 2 | 0 | 0 | 0 |

9   By counting all possible allowed electron states with $n = 1$, $n = 2$ and $n = 3$, show that in each case the number of states with a given value of $n$ is equal to $2n^2$. (This is a general rule.)

10   Which of the following pairs of elements would you expect to be more metallic by virtue of their relative positions in the periodic table?

a   arsenic (As) or bismuth (Bi)

b   lead (Pb) or gold (Au)

c   silicon (Si) or tin (Sn)

d   silicon (Si) or sulfur (S)

e   calcium (Ca) or cesium (Cs)

# molecules and chemical reactions

In the last chapter we examined the concept of the atom and the presently accepted model for its architecture. Most matter can not be understood in terms of a collection of weakly interacting atoms. For example, water is composed exclusively of hydrogen and oxygen atoms, but its properties are surely not the simple sum or average of the properties of these two types of atoms. The reason for this is the unique and specific interactions that occur between different types of atoms. In principle, these interatomic bonds can be explained in terms of one of the four basic interactions discussed in Chapter 1—the electromagnetic force. In practice, the calculations would be impossibly complex for all but the simplest cases. There are just too many particles interacting at the same time. To handle such complexity, chemists have devised models based on averaging or approximating the detailed individual effects. In these models the resulting force holding atoms together is referred to as a *chemical* bond. In water two hydrogen atoms are bound to each oxygen atom forming a triatomic unit called the water *molecule*. It is this structure, rather than the individual atoms that is more meaningfully thought of as the basic unit of water. Sometimes chemical bonding does not result in the formation of individual molecules but rather exceedingly large chains, sheets, or three-dimensional networks of atoms or ions.

Our present understanding of the nature of chemical bonding is another of the triumphs of wave mechanics. As noted above, we are limited to considering the simplest cases if we demand quantitative agreement between theory and experiment. For more complex systems we must presently be satisfied with qualitative, or, at best, semiquantitative agreement. Just as the theory of the many-electron atom is based upon our detailed understanding of the hydrogen atom, so we find it useful to examine the simplest molecules for clues to the explanation of chemical bonding in larger aggregates of atoms.

## 9.1   the hydrogen molecule

The simplest neutral molecule is the hydrogen molecule. It consists of two protons (hydrogen nuclei), and their two electrons, held together by a strong chemical bond. Dalton was unaware that hydrogen gas and several other elemental gases were composed of molecules containing two identical atoms. Once this fact had been established, it remained puzzling to early advocates of the atomic theory that identical atoms should form bonds with one another. In fact, some of the earliest attempts to explain molecule formation were based on the notion that unlike atoms attracted one another, whereas a repulsive force was thought to exist between like atoms.

*Question 9.1*   Can you list other gaseous elements that occur in nature as diatomic species? Which elements exist as monatomic gases?

The fact that the hydrogen molecule exists implies that somehow the system of two protons, a distance $r_{AB}$ apart surrounded by two electrons is more stable than two separated hydrogen atoms. As we have learned, a more stable mechanical system is one with a lower total energy. In Fig. 9-1 the potential energy of two hydrogen atoms is plotted against the separation, $r_{AB}$, between the protons. Indeed there is a drop in energy with a minimum occurring at 0.74 Å. ($1 \text{ Å} = 10^{-10}$ m.) To try to understand why this happens we can examine the potential energy terms involved. Let us label the electrons as 1 and 2 and the protons as $A$ and $B$ as shown in Fig. 9-2. The potential energy due to the attraction at a distance, $r$, between an electron of charge, $-e$, and a proton of charge, $+e$, is given by

$$E_{pot} = \frac{-e^2}{r}.$$ (9-1)

For two electrons (or two protons) a distance, $r$, apart, the potential energy resulting from their mutual repulsion is:

$$E_{pot} = \frac{e^2}{r}.$$ (9-2)

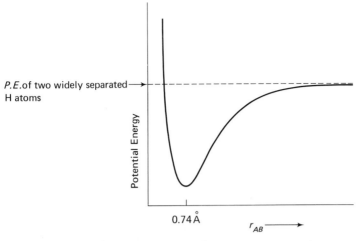

figure 9-1    Potential energy *vs* internuclear separation, $r_{AB}$, for two hydrogen atoms in their lowest energy state forming $H_2$ molecule in its lowest energy state.

The total potential energy of the two hydrogen atoms is actually the sum of the potential energies of the six separate interactions shown in Fig. 9-2.

$$E_{pot\ total} = -\frac{e^2}{r_{1A}} - \frac{e^2}{r_{1B}} - \frac{e^2}{r_{2A}} - \frac{e^2}{r_{2B}} + \frac{e^2}{r_{AB}} + \frac{e^2}{r_{12}}$$

Collecting the common factors ($-e^2$ in the first four terms, and $e^2$ in the fifth

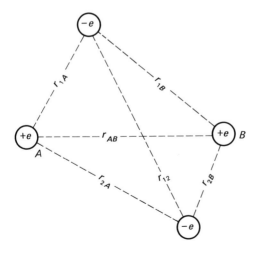

figure 9-2    The possible interactions between protons, with charge ($+e$), and electrons, with charge ($-e$), in the hydrogen molecule.

and sixth terms) yields the expression:

$$E_{pot\ total} = -e^2 \left(\frac{1}{r_{1A}} + \frac{1}{r_{1B}} + \frac{1}{r_{2A}} + \frac{1}{r_{2B}}\right) + e^2 \left(\frac{1}{r_{AB}} + \frac{1}{r_{12}}\right). \quad (9\text{-}3)$$

In Equation 9-3, the first term represents the lowering of the potential energy due to the attraction of the two electrons for the two protons, while the second term indicates an increase in potential energy due to the respective mutual repulsions between the negatively charged pair of electrons, and the positively charged pair of protons.

The expression for the total potential energy of two widely separated hydrogen atoms can be derived from Equation 9-3 in the following manner. In the limiting case of atoms separated by an infinite distance, the interatomic forces (the $1B$ and $2A$ interactions in the first term, and the $AB$ and 12 interactions in the second term) all go to zero. This results in the simplified equation:

$$E_{pot\ total} = -e^2 \left(\frac{1}{r_{1A}} + \frac{1}{r_{2B}}\right) \quad (9\text{-}4)$$

The lowering of the potential energy, as indicated by the shape of the graph in Fig. 9-2 must mean that as the distance between the two nuclei, $r_{AB}$, decreases, the terms which tend to lower the energy of the hydrogen molecule relative to two hydrogen atoms dominate the terms which tend to raise the energy. Some understanding of why this is so can be gained by examining Fig. 9-3. The actual distribution of electron density in the *hydrogen molecule* (as determined from quantum mechanical calculations) is compared to the superposition of the distributions for *two hydrogen atoms* exactly the same distance apart. There is a noticeable shift of the electron density in favor of the region between the nuclei when the diatomic hydrogen molecule is formed from two hydrogen atoms. It is precisely when the electrons are between the

(A)

(B)

figure 9-3   Schematic electron density distributions shown for (A) two overlapping H atoms, (B) the molecule $H_2$. The atoms and the molecule are depicted in their lowest energy states and the internuclear separation distance between nuclei is the same in (A) and (B).

nuclei that the four electron-proton distances $(r_{1A}, r_{1B}, r_{2A}, r_{2B})$ which appear in the denominator of the first (negative) term in Equation 9-3 will be small, resulting in a large negative contribution to the potential energy.

*Question 9.2*   Why do you suppose we have ignored the gravitational forces involved in the formation of the hydrogen molecule?

The origin of the chemical bond is, indeed, the lowering of the potential energy due to a slight shift in the distribution of electrons into the region between the two nuclei.

*Question 9.3*   In terms of the forces involved, show that an electron situated between two nuclei will tend to pull them together, but that an electron outside of this region will tend to push the nuclei apart.

We noted that Fig. 9-1 indicates a minimum in the energy at $r_{AB} = 0.74$ Å. Why does the energy increase when the distance between the nuclei becomes smaller than this? The answer resides in equation 9-3. The increase in energy due to repulsion between the protons is inversely proportional to $r_{AB}$. Quantum mechanical calculations show that it is this term that dominates the expression at very small internuclear distances. Thus a "potential well" results with the minimum at the so-called "equilibrium internuclear distance" of 0.74 Å. The two hydrogen atoms do not remain rigidly fixed—rather, they vibrate around this minimum energy position. Recall the discussion at the end of Section 6.4 about atoms trapped in a potential well.

## 9.2   other molecules

The essential features of bond formation—the lowering of the potential energy due to a shift of electron density into the region between the nuclei, and the resulting potential well—are common to all molecules. Of course, the strengths of the bonds (i.e., the depth of the potential energy minimum), their lengths (i.e., the average internuclear distances), and the actual three dimensional geometry of the molecules are further important details which a sophisticated theory of chemical bonding should explain. Much progress has been made in this direction. Prior to the development of the quantum theory and its successful application to the hydrogen molecule, various regularities in chemical bonding had been observed. The American chemist, Gilbert N. Lewis, published many important papers during the first quarter of this century. In them, he introduced concepts such as *electron sharing,* the *electron pair bond,* and the *octet rule,* for which theoretical justifications were later found.

We have already noted in the case of hydrogen that the pair of electrons is shared, or distributed around both nuclei. Most chemical bonds involve such pairs of electrons. To understand why only two electrons are involved, and not one or three, we need to consider the application of the Pauli Exclusion Principle to molecules. Just as in the case of atoms, this principle states that two electrons with an identical set of quantum numbers will not be present in a molecule. In molecules we again find electrons with an $m_s$ (electron spin) quantum number which can have $+\frac{1}{2}$ and $-\frac{1}{2}$ values—but which has very little effect on the energy of an isolated molecule. The electron distribution shown for hydrogen in Fig. 9-3b is the lowest energy state and will be associated with a set of quantum numbers. Two electrons, differing only in $m_s$ can be accommodated, since one can have an $m_s$ equal to $+\frac{1}{2}$ and the other an $m_s$ equal to $-\frac{1}{2}$. Just as for atoms, the next electron would have to enter a higher state. An electron density map of the next highest energy state of the hydrogen molecule is given in Fig. 9-4. Unlike the lowest energy state, the electron distribution in this state is shifted away from the region between the nuclei, resulting in an *anti-bonding* state. A potential energy diagram analogous to Fig. 9-1, except with the electrons in the anti-bonding state, is given in Fig. 9-5. At all distances the energy is higher than that for the separated atoms and thus a repulsion results. The molecule cannot exist.

The hydrogen molecule has only two electrons and under normal conditions the high energy anti-bonding state is not occupied and has no effect on the bonding. If we consider two helium atoms approaching one another we again find a low energy bonding state available and a higher energy anti-bonding state. Applying the Pauli Exclusion Principle this time, shows that two electrons will go into each state. The repulsion from the two electrons in the anti-bonding state completely cancels the attraction between the two electrons in the bonding state, and there is no net tendency to form a bond. This agrees with the experimental observation that helium exists as a monatomic gas and can not be made to form diatomic molecules.

*Question 9.4*    The Pauli Exclusion Principle also leads to the conclusion that not all collisions between hydrogen atoms will result in the formation of hydrogen molecules.
Explain.

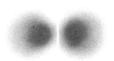

figure 9-4    Schematic electron density distribution in the second lowest potential energy state of the $H_2$ molecule. This state is antibonding. The distribution has been drawn for the case where the nuclei are the same distance apart as those shown in Figure 9-3.

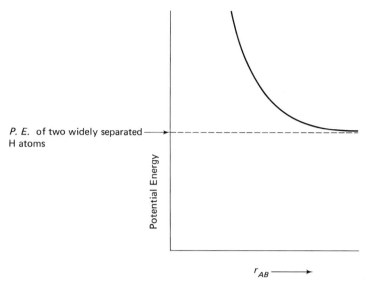

figure 9-5 Potential energy *vs* internuclear separation, $r_{AB}$, for two hydrogen atoms in the next to the lowest energy state (first possible antibonding state).

The octet rule mentioned earlier is of limited general usefulness. The rule originated when Lewis and others noted that many atoms, tended to form bonds until they were surrounded by an outer electron shell, called an "octet," containing eight electrons. This tendency is worth considering in passing, even though there are so many exceptions as to preclude its universal application. In Fig. 9-6 some *electron-dot* pictures of molecules are given to illustrate examples of adherence to the octet rule. In these pictures the symbol for the element is surrounded by dots representing its own *valence* electrons and those which it shares with other atoms. *Valence electrons* are those electrons which participate in bonding. In the examples illustrated, the valance electrons are those having an $n$ quantum number equal to 2. (The reason for this is that

```
                  ..
                 : F :
  ..   ..        .. .. ..           ..
: F : F :      : F : C : F :      H : F :
  ..   ..        .. .. ..           ..
                 : F :
                  ..

   (A)            (B)              (C)
```

```
                                 H
   ..                            .. ..
H : N : H       H : O :       H : C : O : H
   ..             ..             ..
   H              H              H

  (D)            (E)              (F)
```

figure 9-6 Electron dot representation of the valance electron bonding in (A) fluorine molecules, (B) carbon tetrafluoride, (C) hydrogen fluoride, (D) ammonia, (E) water, and (F) methanol.

in the second row of elements in the periodic table the two electrons with $n = 1$ are much closer to the nucleus and therefore will not participate in the electron sharing which leads to bond formation.) Note that if the dots around each atom are counted then all of the atoms other than hydrogen are surrounded by eight electrons (if the electrons between two atoms are considered as belonging to both). The one atom with eight $n = 2$ electrons is neon. This element is one of the very stable, relatively inert gases. Its stability is related to the fact that the next available electronic state is at a much higher energy. Theories of electronic states in molecules justify the stability of the octet of valence electrons on much the same basis.

In approaching the problem of developing theories to explain bonding, modern chemical physicists have adopted two different points of view. The first is to consider the molecule as a unit. The nuclei are considered to be in the positions they occupy in the stable molecule and quantum mechanics is then employed to determine the spatial distributions and energies of the allowed states for the electrons. These states are called orbitals and this method is called the *molecular orbital method*. The second approach consists of considering the formation of bonds between adjacent atoms in a molecule by the bringing together of the individual atoms until the electron distributions appropriate to the isolated atoms overlap and become slightly distorted into the final bonds. This is called the *atomic oribital* or the *valence-bond* approach. Which method is "better" depends on the problem that is being attacked. The molecular orbital method has the advantage that it begins with the correct physical formulation of the problem but the distinct disadvantage that in all but the simplest cases numerous approximations must be introduced in order to make the mathematics manageable. The atomic orbital approach has the advantage that it allows the chemist to attempt to utilize some of his systematic knowledge about the atoms to explain molecular properties. In its least sophisticated form, however, the atomic orbital approach does not explain the many situations in which electrons are shared by more than just a pair of nuclei.

## 9.3   types of bonds

Although we have been considering the general problem of bonding, our discussion has been based on the hydrogen molecule which is a very special example of one type of bond. Bonds between identical atoms are the ideal examples of the so-called *covalent* bonds in which at least one pair of electrons is shared equally by the two atoms involved.

At the other extreme will be bonds between very dissimilar atoms. According to our discussion in Section 8.4 the ionization energy of the chlorine atom is high, whereas that of the potassium atom is quite low. This means that it is much easier to remove electrons from potassium than from chlorine.

Neutral atoms can gain as well as lose electrons. This is due to the fact that although the atom has no net electrical charge, the positive charge is centered in a tiny nucleus while the negative charge is in a diffuse cloud around the nucleus. Therefore, when an extra electron enters the diffuse electron cloud of an atom it will not be completely shielded from the positive nuclear charge. Energy is released when negative ions form as a result of the attraction between neutral atoms and electrons. The quantitative measure of this energy is called the *electron affinity*. The values of the electron affinities of a few atoms are given in Table 9-1. We see that potassium and chlorine also differ greatly in their electron affinities.

*Question 9.5*    From handbook data, and the data in Table 9-1, how does electron affinity vary with the position of the atom in the periodic table? Can you explain this trend on the basis of the quantum model of the atom?

When potassium and chlorine interact, the result is *not* a diatomic molecule with a pair of electrons shared equally by the two nuclei. Instead, the product is a crystalline solid called a salt. In this case a white crystalline salt forms, consisting of negative chloride ions (the Cl atom plus one electron) and positive

table 9.1    electron affinities of gaseous atoms.

| Atom | Affinity (kcal/mole)* |
|------|-----------------------|
| H    | 17.3                  |
| C    | 29                    |
| O    | 34                    |
| F    | 79.5                  |
| Na   | 19†                   |
| Si   | 32†                   |
| S    | 47                    |
| Cl   | 83.4                  |
| K    | 16†                   |
| Br   | 77.3                  |
| I    | 70.5                  |

° One kcal = 1000 calories
   One mole = $6.02 \times 10^{23}$ atoms.
(See the discussion of the mole in Section 9.7.)
   † Estimated from calculations rather than based on experimental measurement.

potassium ions (the K atom minus one electron). These ions are arranged alternately in a three dimensional array as shown in Fig. 9-7.

If we look into the detailed energetics of the problem we find that although the electron affinity of chlorine is high, the energy released when the negative ion is formed is not enough to remove an electron from potassium. Why then does the reaction proceed? When the resulting positive and negative ions come together additional energy is released. Thus, when an electron is transferred from an atom with a low ionization energy (like potassium) to an atom with a high electron affinity (like chlorine) the resulting attraction between the oppositely charged ions lowers the potential energy sufficiently to make the entire process energetically favorable.

In general, the formation of such ionic compounds can occur between many pairs of elements if one is a metal whose atoms have a low ionization energy and electron affinity, and the other is a non-metal whose atoms have high values for both of these properties.

*Question 9.6*    Why wouldn't you expect ionic bonds to form between two metals or two non-metals?

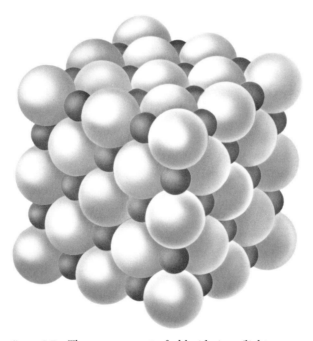

figure 9-7    The arrangement of chloride ions (light spheres) and potassium ions (dark spheres) in a crystal of potassium chloride. This figure is from an early paper by William Barlow. (From *General Chemistry*, Third Ed., by L. Pauling. Copyright © 1970. W. H. Freeman and Co.)

figure 9-8 Representation of equal sharing of valence electrons
in the hydrogen molecule (left) and the fluorine molecule (right),
and the unequal sharing in hydrogen fluoride (center).

The virtually complete transfer of one or more electrons from one atom
to another that is involved in *ionic* bonding is the opposite extreme from the
equal sharing of electrons in the pure covalent bond formed by two identical
atoms. Between these two extremes, there exists a continuum of bond types
that can be described either as partly ionic and partly covalent or as involving
the unequal sharing of pairs of electrons. A schematic representation of
unequal as compared to equal sharing in the H—H, H—F and F—F bonds
is shown in Fig. 9-8.

The unequal sharing of electrons results in a shift of the negative charge
towards one atom with the other atom becoming slightly positively charged,
as a result. The physical term for such a neutral object with an internal charge
separation is a *dipole*. The magnitude of the dipole, called the *dipole moment*,
is given by the product of the amount of charge separated and the distance
between the charges. The more dissimilar that two atoms in a diatomic
(two-atom) molecule are, the larger the charge separation will be. Some dipole
moments of diatomic molecules are given in Table 9-2.

Thus far we have considered bonds that result from the localized sharing
of pairs of electrons between two adjacent atoms. In some cases, molecular

table 9.2 dipole moments of some diatomic
molecules.

| Molecule | Dipole Moment (D) $(D = $ Debye Units $= 10^{-18}$ esu $\cdot$ cm)* |
|---|---|
| HF | 1.91 |
| HCl | 1.07 |
| HI | 0.38 |
| LiF | 6.28 |
| LiH | 5.88 |
| ClF | 0.88 |
| BrF | 1.29 |
| NO | 0.15 |
| SrO | 8.90 |

*One esu is a unit of electrical charge equal to
$3.3 \times 10^{-10}$ coulombs.

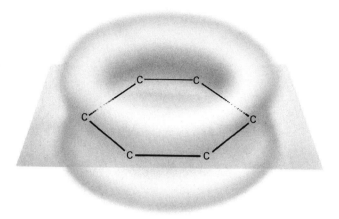

figure 9-9   Delocalized bonding in benzene. In addition to
localized bonds between the six carbon atoms in the
planar hexagonal ring, there are six electrons in three
delocalized orbitals each of which extends over the entire
molecule above and below the plane of carbon nuclei.

orbitals can extend over several atoms. The electrons that occupy these orbitals
are said to be *delocalized*. For example, in the benzene molecule, there are
six carbon atoms in a hexagonal arrangement. Each of these carbon atoms
forms three localized bonds by sharing pairs of electrons with each of its two
neighbors and with a hydrogen atom. In addition, six electrons are in molecular
orbitals which are delocalized over the entire six carbon atom ring as shown
in Fig. 9-9.

## 9.4   polar and nonpolar molecules

Any diatomic molecule, other than those involving two identical atoms, will
have a bond that is at least slightly polar, and which will give rise to a dipole
moment. For a polyatomic molecule we must consider the geometry as well
as the polarity of individual bonds in order to determine whether a dipole
moment will result. As an illustration let us consider the two molecules, carbon
dioxide and water, as shown in Fig. 9-10. Both molecules consist of a central
atom of one kind bound to two identical atoms of another kind. In each case
the two bonds are polar. In carbon dioxide the linear arrangement of the bonds
results in an exact cancellation of the two bond dipoles and the molecule has
no dipole moment. The angular geometry of the water molecule, on the other
hand, results in a rather large net dipole moment.

*Question 9.7*   $BF_3$ and $NF_3$ molecules each consist of an equilateral triangle of fluorine
atoms which form three polar bonds to a different atom. In $BF_3$ the B
atom is in the center of the triangle (forming a planar molecule), whereas

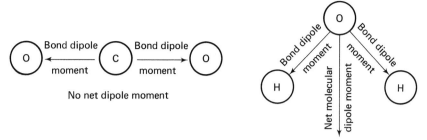

figure 9-10   In the linear carbon dioxide molecules (left) the bond dipoles cancel. In non-linear $H_2O$ molecules (right) the bond dipoles add to give a net molecular dipole moment.

in $NF_3$ the N atom is above (or below) the center of the $F_3$ plane. Explain why $BF_3$ has no dipole moment while $NF_3$ does.

When a molecule with a dipole moment is placed between two oppositely charged plates it tends to align itself with its negative end toward the positive plate and with its positive end toward the negative plate as shown in Fig. 9-11. For a laboratory sample containing a very large number of molecules, the extent to which this alignment occurs can be used to determine the magnitude of the molecular dipole moment. The dipole moments of several diatomic molecules are given in Table 9-2.

## 9.5   chemical reactions and the chemical equation

The burning of rocket fuel, the rusting of unprotected iron, and the digestion of food by the body are all examples of processes in which matter undergoes an obvious change. All macroscopic changes reflect rearrangements on the molecular level in which new chemical bonds are formed and/or old bonds are broken. The term used to refer to such molecular reorganization is *chemical reaction*.

Some chemical reactions occur extremely rapidly. The explosive reaction that results when a spark is introduced into a mixture of hydrogen and oxygen gases is an example. Other chemical reactions can be imperceptibly slow; for example, the undisturbed reaction between hydrogen and oxygen at room temperature. Almost all of the scientific investigation done by chemists—and also much of the work of biologists, geologists, and space scientists—is directly or indirectly involved with the study of chemical reactions. They are seeking to discover the answers to such questions as: Why do certain molecular rearrangements occur while others do not? Why do specific reactions occur at particular rates? What is the detailed mechanism by which the molecular

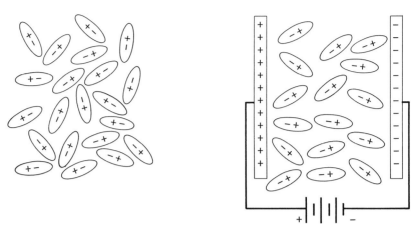

figure 9-11    Polar molecules in a liquid or gas will have no net orientation of their dipole moments. When the sample is placed between two oppositely charged plates, the dipoles will tend to line up as shown above.

restructuring takes place? What conditions of temperature, pressure, or contaminants will favor a particular reaction and suppress another? How much energy is required or released when a reaction occurs?

A concise method of indicating what net changes occur in a chemical reaction has been developed. It takes the form of an algebraic equation. Since ordinary chemical changes do not involve the destruction of atoms, we can equate the numbers of each type of atom in the products of a reaction to the numbers of each type initially present in the unreacted substances. An example of such a *chemical equation* is the one for the formation of water from hydrogen and oxygen:

$$2 \, H_2 + O_2 = 2 \, H_2O \qquad (9\text{-}5)$$

Molecules are represented by stringing together the letter symbols for their constituent atomic elements with a numerical subscript following the atomic symbol to indicate how many atoms of that type are involved. Thus, hydrogen and oxygen are written as $H_2$ and $O_2$ since they are diatomic molecules. Water, consisting of two atoms of hydrogen and one of oxygen, is written $H_2O$. (The subscript is omitted whenever only one atom of a particular element is present.) In the equation, the coefficients preceding the formulas for the molecules—2 for $H_2$, 1 (again the 1's are not written) for $O_2$ and 2 for $H_2O$—indicate the relative numbers of each of the molecules involved in the reaction. The requirement that equal numbers of hydrogen atoms (4 in this case), and oxygen atoms (2 in this case), appear on each side of the equal sign in equation 9-5 is fulfilled. This could also be accomplished by writing:

$$4 \, H_2 + 2 \, O_2 = 4 \, H_2O \qquad (9\text{-}6)$$

or,

$$H_2 + \tfrac{1}{2} O_2 = H_2O. \tag{9-7}$$

Although it is common practice to use the equation with the smallest integral coefficients—equation 9-5 in this instance—equations 9-6 and 9-7 are also perfectly respectable balanced equations. On the other hand the equation written:

$$H_2 + O_2 = H_2O \tag{9-8}$$

is not acceptable since different numbers of oxygen atoms appear on the two sides of the equal sign. The process of adjusting the coefficients so that atoms are conserved is called *balancing the equation*.

*Question 9.8*   Examine the following three chemical equations to see if each is balanced. Balance any unbalanced equations.

$$2\,C_8H_{18} + 25\,O_2 = 16\,CO_2 + 18\,H_2O$$

$$CH_2O + O_2 = CO_2 + H_2O$$

$$NH_3 + O_2 = N_2 + H_2O$$

These equations represent the burning of octane $(C_8H_{18})$, formaldehyde $(CH_2O)$, and ammonia $(NH_3)$. The products include carbon dioxide $(CO_2)$ and nitrogen $(N_2)$.

  The chemical equations considered thus far have involved substances composed only of discrete molecules. Some compounds, such as salts, are not molecular in nature. A crystal of potassium chloride (Fig. 9-7) consists of positively charged potassium ions $(K^+)$ and negatively charged chloride ions $(Cl^-)$ arranged in a repetitive pattern. Although there are equal numbers of $K^+$ and $Cl^-$ ions in any crystal, they do not form molecular aggregates such as $KCl$, $K_2Cl_2$ or $K_3Cl_3$. If you wish, you can think of each little crystal as a giant molecule, but this requires a change in our previous notion of a molecule as an aggregate of a *specific* number of atoms. It is more useful to retain the more limited definition of molecules and to exclude ionic solids (such as salts) from the class of molecular substances. Covalently bonded substances can also be non-molecular. Polymers, of which polyethylene, nylon, and dacron are three familiar examples, consist of one or more atomic group-

ings (such as H—C—H in the case of polyethylene) covalently bonded together

to form molecular strands (or sheets) of variable lengths or sizes. Again we can not write a unique molecular formula. When writing reactions involving compounds of definite atomic composition which are not molecular, the chemist uses the so-called *simplest formula*. This formula expresses the ratios of the atoms present in the compound by using the smallest whole number ratio. For potassium chloride this would be KCl. Thus the reaction for the formation of potassium chloride from potassium metal and chlorine gas is written:

$$2 \text{ K} + \text{Cl}_2 = 2 \text{ KCl} \tag{9-9}$$

*Question 9.9*    It is also possible to write simplest formulas for compounds which are molecular. What would be the simplest formulas for benzene (molecular formula: $C_6H_6$), and glucose (molecular formula: $C_6H_{12}O_6$)?

Another consideration in formula writing is that the same compound may have a different formula under different conditions. Ionic solids when dissolved in water are completely dissociated into their component ions. Sodium chloride solutions consist of sodium ions ($Na^+$), chloride ions ($Cl^-$), and water molecules. Silver nitrate, another ionic solid, dissociates into silver ($Ag^+$) and nitrate ($NO_3^-$) ions when dissolved in water. The ionic solid, silver chloride (AgCl) has a very low solubility in water. When solutions of silver nitrate and sodium chloride are mixed the free $Ag^+$ and $Cl^-$ ions will combine to form solid AgCl which will precipitate out. (When precipitation occurs it is indicated by the subscript *s* next to the symbol for the chemical that is precipitated.) How should we write the chemical equation for this reaction? If we write:

$$\text{AgNO}_3 + \text{NaCl} = \text{AgCl}_{(s)} + \text{NaNO}_3 \tag{9-10}$$

we would be misrepresenting the facts, since $AgNO_3$, NaCl and $NaNO_3$ were not present before or after the reaction. Before the reaction we have $Ag^+$, $NO_3^-$, $Na^+$, $Cl^-$, and $H_2O$ present and after the reaction we have AgCl, $Na^+$, $NO_3^-$, and $H_2O$ present. The most meaningful equation for the *net* reaction is written:

$$\text{Ag}^+ + \text{Cl}^- = \text{AgCl}_{(s)} \tag{9-11}$$

Note that $Na^+$, $NO_3^-$, and $H_2O$ do not appear in this net equation. This is as it should be because, although they were present throughout, the chemical change that occurred did not involve any of these species.

*Question 9.10*    What would be the net equation for the chemical reaction between solid zinc metal and hydrochloric acid that produces hydrogen gas and a solution of zinc chloride? (Hydrochloric acid consists of H⁺ and Cl⁻ ions dissolved in water. Zinc chloride is completely dissociated into $Zn^{2+}$ and Cl⁻ ions in water solutions.)

Frequently subscripts following the formula of a compound are used to indicate whether it is in the solid ($s$), gaseous ($g$), or liquid ($l$) form. The absence of a subscript after a particular species generally indicates that it is present in a water (aqueous) solution. Where there might be some ambiguity the subscript $aq$ is used to indicate an aqueous solution. Equations 9-5, 9-9, and 9-11 have been rewritten to illustrate the use of these subscripts as:

$$2 H_{2(g)} + O_{2(g)} = 2 H_2O_{(l)} \qquad (9\text{-}5a)$$

$$2 K_{(s)} + Cl_{2(g)} = 2 KCl_{(s)} \qquad (9\text{-}9a)$$

$$Ag^+_{(aq)} + Cl^-_{(aq)} = AgCl_{(s)} \qquad (9\text{-}11a)$$

## 9.6   atomic and molecular weights

In Chapter 8 it is stated that observations of the periodic properties of the elements resulted in the preparation of tables of the elements that were ordered according to the masses of the atoms. How were Nineteenth century scientists able to measure the masses of things as small as atoms? In fact, they were not able to determine individual atomic masses. However, they were able to deduce the *relative* masses of the atoms. In order to do this they only had to know the formula (number of atoms of each kind) in a compound and the percentage by weight of each element in the compound.

For example, suppose a compound is known to consist of molecules made up of one atom of carbon and one atom of oxygen. This compound is found to be 42.86 percent carbon and 57.14 percent oxygen by weight. A 100 gram sample of the substance will contain 42.86 grams of carbon and 57.14 grams of oxygen. Without knowing how many total atoms are present we know that there are *equal* numbers of carbon and oxygen atoms. Therefore the ratio 42.86:57.14 must be the relative masses of individual atoms of carbon and oxygen!

As a second example we can take water. The percentage composition by weight is 11.11 percent hydrogen and 88.89 percent oxygen. A 100 gram sample will be composed of 11.11 grams of hydrogen and 88.89 grams of oxygen. There will be twice as many hydrogen atoms in the sample as oxygen atoms since the formula is $H_2O$.[1] This means that 88.89:11.11 will represent

---

1. In fact, early chemists thought water had the formula HO. Such errors led to much confusion and many arguments about the correct assignment of relative atomic masses.

the ratio of the mass of one oxygen atom to *two* hydrogen atoms. The ratio of one oxygen atom to *one* hydrogen atom will be twice as large (2 × 88.89 : 11.11).

*Question 9.11*    On the basis of the two examples given above determine the relative masses of carbon and hydrogen atoms.

It is possible to define an "atomic mass system" by arbitrarily choosing one element as a standard and defining the *atomic mass unit* in terms of the atoms of this element. (A particular isotope of carbon is now the universally accepted standard, and its mass is set equal to exactly 12 atomic mass units (AMU). The number 12 is chosen because it makes the mass of the lightest atom 1.0 AMU, and it results in nearly integral masses for many of the other light elements.) Until techniques, such as x-ray crystalography and mass spectroscopy, were developed for measuring the mass of an individual atom in grams, the conversion factor between the AMU and the gram was undetermined. Today we know that $1.00 \text{ AMU} = 1.66 \times 10^{-24}$ gram.

*Question 9.12*    What is the mass of an atom of carbon in grams?

The mass of an atom of a particular element in AMU is commonly called the *atomic weight* of that element. Thus hydrogen, carbon, and oxygen have atomic weights of 1.0, 12.0, and 16.0 respectively. (The use of the term weight instead of mass is unfortunate but scientists are occasionally guilty of sloppy or inconsistent terminology due to historical precedents and usages of terms.)

*Question 9.13*    Do these atomic weights jibe with the relative masses of the hydrogen, carbon, and oxygen atoms given above?

Table 9-3 lists the names, atomic numbers, and atomic weights of all of the elements. The atomic weight is actually the *average* mass in AMU of the atoms of an element. Most elements are now known to be composed of two or more types of atoms with the same chemical properties but different masses. Atoms of the same element with different masses are called *isotopes.*

The bulk of an atom's mass results from the particles in the nucleus known as protons and neutrons. The proton carries a positive charge exactly equal in magnitude to the negative charge on the electron, while the neutron carries no charge. Both of these particles have masses equal to 1.0 AMU. All of the isotopes of a particular element will have the same number of protons (equal

table 9.3 atomic symbols, numbers, and average weights in AMU of the elements arranged in alphabetical order

| Element | Symbol | Atomic Number | Atomic* Weight | Element | Symbol | Atomic Number | Atomic* Weight |
|---------|--------|---------------|----------------|---------|--------|---------------|----------------|
| Actinium | Ac | 89 | (227) | Mercury | Hg | 80 | 200.59 |
| Aluminum | Al | 13 | 26.98 | Molybdenum | Mo | 42 | 95.94 |
| Americium | Am | 95 | (243) | Neodymium | Nd | 60 | 144.24 |
| Antimony | Sb | 51 | 121.75 | Neon | Ne | 10 | 20.183 |
| Argon | Ar | 18 | 39.948 | Neptunium | Np | 93 | (237) |
| Arsenic | As | 33 | 74.922 | Nickel | Ni | 28 | 58.71 |
| Astatine | At | 85 | (210) | Niobium | Nb | 41 | 92.906 |
| Barium | Ba | 56 | 137.34 | Nitrogen | N | 7 | 14.007 |
| Berkelium | Bk | 97 | (249) | Nobelium | No | 102 | (255) |
| Beryllium | Be | 4 | 9.012 | Osmium | Os | 76 | 190.2 |
| Bismuth | Bi | 83 | 208.98 | Oxygen | O | 8 | 15.9994 |
| Boron | B | 5 | 10.811 | Palladium | Pd | 46 | 106.4 |
| Bromine | Br | 35 | 79.909 | Phosphorus | P | 15 | 30.974 |
| Cadmium | Cd | 48 | 112.40 | Platinum | Pt | 78 | 195.09 |
| Calcium | Ca | 20 | 40.08 | Plutonium | Pu | 94 | (242) |
| Californium | Cf | 98 | (251) | Polonium | Po | 84 | (210) |
| Carbon | C | 6 | 12.011 | Potassium | K | 19 | 39.102 |
| Cerium | Ce | 58 | 140.12 | Praseodymium | Pr | 59 | 140.91 |
| Cesium | Cs | 55 | 132.91 | Promethium | Pm | 61 | (147) |
| Chlorine | Cl | 17 | 35.453 | Protactinium | Pa | 91 | (231) |
| Chromium | Cr | 24 | 51.996 | Radium | Ra | 88 | (226) |
| Cobalt | Co | 27 | 58.933 | Radon | Rn | 86 | (222) |
| Copper | Cu | 29 | 63.54 | Rhenium | Re | 75 | 186.2 |
| Curium | Cm | 96 | (247) | Rhodium | Rh | 45 | 102.91 |
| Dysprosium | Dy | 66 | 162.50 | Rubidium | Rb | 37 | 85.47 |
| Einsteinium | Es | 99 | (254) | Ruthenium | Ru | 44 | 101.07 |
| Erbium | Er | 68 | 167.26 | Samarium | Sm | 62 | 150.35 |
| Europium | Eu | 63 | 151.96 | Scandium | Sc | 21 | 44.96 |
| Fermium | Fm | 100 | (253) | Selenium | Se | 34 | 78.96 |
| Fluorine | F | 9 | 19.00 | Silicon | Si | 14 | 28.09 |
| Francium | Fr | 87 | (223) | Silver | Ag | 47 | 107.870 |
| Gadolinium | Gd | 64 | 157.25 | Sodium | Na | 11 | 22.9898 |
| Gallium | Ga | 31 | 69.72 | Strontium | Sr | 38 | 87.62 |
| Germanium | Ge | 32 | 72.59 | Sulfur | S | 16 | 32.064 |
| Gold | Au | 79 | 196.97 | Tantalum | Ta | 73 | 180.95 |
| Hafnium | Hf | 72 | 178.49 | Technetium | Tc | 43 | (99) |
| Helium | He | 2 | 4.0026 | Tellurium | Te | 52 | 127.60 |
| Holmium | Ho | 67 | 164.93 | Terbium | Tb | 65 | 158.92 |
| Hydrogen | H | 1 | 1.0080 | Thallium | Tl | 81 | 204.37 |
| Indium | In | 49 | 114.82 | Thorium | Th | 90 | 232.04 |
| Iodine | I | 53 | 126.90 | Thulium | Tm | 69 | 168.93 |
| Iridium | Ir | 77 | 192.2 | Tin | Sn | 50 | 118.69 |
| Iron | Fe | 26 | 55.847 | Titanium | Ti | 22 | 47.90 |
| Krypton | Kr | 36 | 83.80 | Tungsten | W | 74 | 183.85 |
| Lanthanum | La | 57 | 138.91 | Uranium | U | 92 | 238.03 |
| Lawrencium | Lw | 103 | (257) | Vanadium | V | 23 | 50.94 |
| Lead | Pb | 82 | 207.19 | Xenon | Xe | 54 | 131.30 |
| Lithium | Li | 3 | 6.939 | Ytterbium | Yb | 70 | 173.04 |
| Lutetium | Lu | 71 | 174.97 | Yttrium | Y | 39 | 88.905 |
| Magnesium | Mg | 12 | 24.312 | Zinc | Zn | 30 | 65.37 |
| Manganese | Mn | 25 | 54.938 | Zirconium | Zr | 40 | 91.22 |
| Mendelevium | Md | 101 | (258) | | | | |

* The elements whose atomic weights appear in parenthesis do not occur in nature. The atomic weight listed for these elements is that of the most common man-made isotope.

to the element's atomic number) in their nuclei, but will differ in the number of neutrons. For example, naturally occurring chlorine consists of two isotopes. One isotope contains 18 neutrons and 17 protons and the other contains 20 neutrons and 17 protons. Since the mass of an electron is only 1/1840 that of a proton or a neutron, the atomic weight of particular isotopes will be very nearly equal to the sum of the number of protons and neutrons in its nucleus. Thus the two isotopes of chlorine have atomic weights of 35.0 and 37.0 AMU. The percentages of these two isotopes in any naturally occurring sample of chlorine is found to be 75.4% and 24.6% respectively. The average atomic weight of chlorine is therefore equal to

$$\frac{(75.4 \times 35.0 + 24.6 \times 37.0)}{100} \text{ or } 35.5 \text{ AMU.}$$

Note that in Table 9-3 the atomic mass of carbon is given as 12.01, not as 12.00. This is because the AMU is defined in terms of the particular isotope of carbon that has 6 protons and 6 neutrons. This isotope has, by definition, an atomic weight of exactly 12.000 . . . . The carbon that occurs in nature is a mixture of 98.9% of this isotope with 1.1% of an isotope having 6 protons and 7 neutrons.

*Question 9.14*   The fact that many of the light elements have nearly integral atomic weights is a reflection of the fact that they consist almost exclusively of a single isotope.
Explain.

The mass of a molecule is simply equal to the sum of the masses of its constituent atoms. If we add up the atomic weights of all of the atoms in a molecule the result will be the *molecular weight*. Thus water, $H_2O$, has a molecular weight of 18.0; $(2 \times 1.0 + 16.0 = 18.0)$. For those substances which do not consist of molecules, but for which we can write simplest formulas, we can refer to the *formula weight*. Thus the formula weight of NaCl is 58.5; $(23.0 + 35.5 = 58.5)$.

## 9.7   chemical arithmetic

Of frequent practical concern both to the researcher in his laboratory and to the industrial scientist or engineer are the quantitative aspects of chemical change. How much X can be made from a given amount of Y and Z? The balanced net chemical equation is the basis for answering such questions. Equation 9-9 reads: "Two atoms of potassium will combine with one molecule of chlorine to produce 2 formula units of potassium chloride." (In cases where we are dealing with non-molecular compounds the set of atoms indicated by the simplest formula will be called a *formula unit.*) This of course means that if we start with six atoms of potassium and three molecules of chlorine we can produce six formula units of KCl.

Suppose we have 1000 atoms of potassium and 600 molecules of chlorine. How many formula units of potassium chloride can be produced? The chemical equation indicates that we need twice as many K atoms as $Cl_2$ molecules. We have less K than the amount required to combine with all the $Cl_2$. The first reactant to be completely consumed is called the *limiting* reactant. It will limit the amount of product formed. When all the K is gone the reaction will cease. At this point 500 molecules of $Cl_2$ will have been consumed and 1000 KCl formula units will have been produced. The remaining 100 $Cl_2$ molecules will be left over.

The steps involved in determining how much of any product can be formed in a chemical reaction are:

1   Write the balanced net equation for the process.
2   Determine the limiting reactant.
3   Use the equation to determine how much product can be formed from the given amount of limiting reactant.

*Question 9.15*   How many $H_2O$ molecules can be formed from 600 molecules of $H_2$ and 800 molecules of $O_2$?

In actual practice we do not measure reactants and products in terms of the number of atoms and molecules involved, but rather in terms of grams or kilograms. In the previous section it was stated that 1 AMU = $1.66 \times 10^{-24}$ grams. This means that 1.00 gram of carbon atoms, which has an average atomic weight of 12.01, will contain

$$\frac{1.00}{(12.01)(1.66 \times 10^{-24})} = \left(\frac{1.00}{12.01}\right)(6.02 \times 10^{23})$$

carbon atoms. If we have 12.01 grams of carbon the number of atoms present will be

$$\left[\frac{(12.01)(1.00)}{12.01}\right](6.02 \times 10^{23}) = 6.02 \times 10^{23}$$

atoms. Similarly for any atom (or molecule) if we have a sample that has a mass in grams exactly equal to its atomic (or molecular) weight, the number of atoms (or molecules) respectively present will be $6.02 \times 10^{23}$. This number is called Avogadro's number in honor of the Italian scientist Amadeo Avogadro, who was the first to propose how to obtain equal numbers of molecules of different substances. (Avogadro, who died in 1856, never knew the actual magnitude of "his" number.) The amount of any substance containing Avogadro's number of units is called a *mole*. Thus a mole of hydrogen atoms will have a mass of 1.01 grams and will contain $6.02 \times 10^{23}$ H atoms. A mole of hydrogen molecules (sometimes referred to simply as a mole of hydrogen) will have a mass of 2.02 grams and will contain $6.02 \times 10^{23}$ $H_2$ molecules. A mole of sodium chloride will have a mass of 58.5 grams and will contain $6.02 \times 10^{23}$ formula units of NaCl. We may thus define a mole of a substance as the number of grams equal to the molecular weight of the substance.

Since a mole of any substance always involves the same number of atoms, molecules, or formula units we can use this macroscopic unit in working chemical problems. For example, we can read Equation 9-12 as: "Two moles of Na will combine with one mole of $Cl_2$ to produce two moles of NaCl."

$$2 \text{ Na} + Cl_2 = 2 \text{ NaCl} \tag{9-12}$$

Suppose we react 10.0 grams of sodium with 15.0 grams of chlorine. How many grams of NaCl will be formed?

$$10.0 \text{ grams of Na} = \frac{10.0}{23.0} = 0.435 \text{ moles Na}$$

$$15.0 \text{ grams of } Cl_2 = \frac{15.0}{(2)(35.5)} = 0.211 \text{ moles of } Cl_2$$

The molar ratio of Na to $Cl_2$ is larger than the 2:1 ratio indicated by the balanced equation for the reaction. There is more sodium present than necessary to use up all the chlorine. Therefore $Cl_2$ will be the limiting reactant. The 0.211 moles of $Cl_2$ will be completely consumed, using up 0.422 moles of the Na, and 0.422 moles of NaCl will be produced.

$$0.422 \text{ moles of NaCl} = (0.422)(23.0 + 35.5)$$

$$= 24.7 \text{ grams of NaCl}$$

Working with moles is very little different from working with atoms or molecules. The only additional steps involve the use of the relationship,

$$\text{moles} = \frac{\text{grams}}{\text{atomic (or molecular) weight}}$$

to convert mass to moles and vice versa.

*Question 9.16*    How many grams of $H_2O$ can be formed from 10.0 grams of $H_2$ and 30.0 grams of $O_2$?

When liquid reactants are involved, it is the common practice to state the quantity of reactant in terms of the volume of the liquid. Working problems involving liquids will frequently require converting volumes to grams. This requires a knowledge of the *density* of the liquid. This is merely the ratio of the mass to the volume:

$$\text{grams of liquid} = (\text{density})(\text{volume})$$

*Question 9.17*  Ethyl alcohol, $C_2H_6O$, has a density of 0.789 grams/cm$^3$ at room temperature. How many grams of $O_2$ will be necessary to burn completely 10.0 cm$^3$ of $C_2H_6O$? The equation for the reaction is:

$$C_2H_6O + 3\,O_2 = 2\,CO_2 + 3\,H_2O$$

## summary

We examined the nature of the chemical bond in the hydrogen molecule since, as in the case of atoms, this simplest molecule provides the basis for understanding more complicated species. We discovered that those quantized states of molecules for which the electronic distribution is concentrated between the nuclei are the bonding states, as opposed to anti-bonding states in which the density of electrons is greater outside the inter-nuclear region. The Pauli Exclusion Principle as applied to molecules explains why bonds are only formed between atoms when electrons can be accommodated in bonding molecular states. The observed "octet rule" for electron distribution in molecules is rationalized in terms of the particularly stable electronic configurations which this rule predicts.

In examining other molecules we find a continuum of bonding situations from the completely ionic bond, in which electrons are transferred between atoms, to the completely covalent bond involving equal sharing of electrons. The existence of unequal sharing of electrons in bonds between dissimilar atoms is found to result in bond polarity which often leads to a net charge separation (dipole moment) for the overall molecule.

The shorthand notation by which we write chemical equations to represent bonding rearrangements that occur during chemical reactions is explained. The net equation in which only the actual changes that occur is shown to be the most useful form for these equations.

The basis for the atomic "weight" system is examined. The mass of a particular isotope of carbon is defined as twelve atomic mass units (AMU) and the masses of all other atoms are measured relative to this standard.

Using chemical equations and the atomic "weight" system we showed some simple chemical arithmetic. Since the AMU is much too small for use in practical problems (1 AMU = $1.66 \times 10^{-24}$ g) a new unit is required. We therefore defined a new unit for use in such problems: the number of grams equal to the molecular weight of a substance as a *mole* of that substance. It is observed that a mole of any material will always contain the same number of molecules ($6.02 \times 10^{23}$), and therefore this unit can replace the molecule as the basis for chemical computations.

Answers to
Questions

9.1  Nitrogen, oxygen, fluorine, and chlorine are the four other gaseous elements that occur as diatomic molecules. The other naturally occurring gaseous elements, helium, neon, argon, krypton, xenon, and radon are all members of the so-called noble gas family and occur in nature as uncombined atoms.

9.2  The gravitational attraction is negligible compared to the electromagnetic forces involved in the interactions between atoms.

9.3  Consider an electron between two nuclei; the attractive force which results between the nuclei and the electron will partially cancel the repulsion of the like-charged nuclei. If the electron is outside the region between the nuclei it will still be attracted by both centers of positive charge, but it will attract the nearer nucleus more strongly and, rather than cancelling the internuclear repulsion, it will enhance the forces tending to separate the nuclei.

9.4  As we have seen, a hydrogen molecule will form only if both electrons occupy the lowest bonding energy state. For this to happen the two electrons must have opposite "spins." If the two approaching hydrogen atoms both have electrons with the same spin quantum number, then one of the electrons will end up in the next to the lowest (anti-bonding) energy state, and a stable molecule will not form.

9.5  Electron affinity increases in going from left to right in the periodic table until the last column is reached, at which point it drops to a much lower value than that for the previous element. Going down a column in the periodic table results in a decrease in electron affinity. This behavior can be understood if we recall the ordering and separation of allowed quantized energy states for electrons. For example, as we proceed from lithium to fluorine we will always be adding the extra electron in an $n = 2$ state, but the nuclear charge will be increasing. This increased nuclear charge will be only partly shielded by the electrons and as a result the attraction for an additional electron will be getting greater. When we come to neon we have no more room in the $n = 2$ "shell" and the extra electron must be accommodated in an $n = 3$ state—considerably higher in energy and further from the nucleus. This outer electron would be much less tightly bound. Going down a column means that we are considering electrons in higher energy states, thus the decrease in electron attraction.

9.6  For similar elements we have ionization energies and electron affinities which are either both relatively low (metals) or high (non-metals). In these cases the energy released upon adding an electron to one atom is so much less than the energy required to remove an electron from the other atom that the additional energy release that would result from attraction of the charged ions formed is not enough to make up the difference.

9.7  In $BF_3$ the bond dipoles between the boron and the three fluorines are all in the same plane with 120° angles between them. In such an arrangement each of these dipoles will be exactly cancelled by the effect of the other two. In $NF_3$ the three bond dipoles are not in a plane. The negative fluorines form a plane of negative charge with the relatively positively charged N atom either above or below it. A net dipole pointing from the center of the fluorine triangle to the nitrogen atom will result.

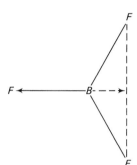

figure 9-12

9.8  The first two equations are balanced. The third one can be balanced by first noting that only one N appears on the left, while there are two on the right. We must therefore multiply $NH_3$ by two. We now have six H's on the left and only two on the right. This can be corrected by multiplying $H_2O$ by three. We now find only two O's on the left and three on the right. A balanced equation can now be written if we multiply $O_2$ by $\frac{3}{2}$.

$$2\,NH_3 + \tfrac{3}{2}\,O_2 = N_2 + 3\,H_2O$$

If an equation without fractions is preferred we need only multiply the whole thing by two:

$$4\,NH_3 + 3\,O_2 = 2\,N_2 + 6\,H_2O$$

9.9  The simplest formulas would be: CH for benzene, and $CH_2O$ for glucose.

9.10  The net equation would be:

$$Zn + 2\,H^+ = H_2 + Zn^{2+}$$

9.11

$$\frac{\text{mass of C atom}}{\text{mass of H atom}} = \left(\frac{\text{mass of C atom}}{\text{mass of O atom}}\right)\left(\frac{\text{mass of O atom}}{\text{mass of H atom}}\right)$$

$$= \left(\frac{42.86}{57.14}\right)\left(\frac{2 \times 88.89}{11.11}\right) = \frac{12.00}{1.00}$$

9.12  Mass of one carbon atom $= (12)(1.66 \times 10^{-24}) = 1.99 \times 10^{-23}$ g.

9.13   Yes:

$$\frac{42.86}{57.14} = \frac{12}{16}, \qquad \frac{(2)(88.92)}{11.11} = \frac{16}{1}$$

9.14   If a light element consists exclusively of one isotope its atomic weight in AMU will be approximately equal to the sum of the number of neutrons and protons in its nucleus. Thus the fluorine found in nature contains atoms with 9 protons and 10 neutrons exclusively, and has an atomic weight of 19.0.

9.15   $2 H_2 + O_2 = 2 H_2O$. The limiting reactant is $H_2$: 600 molecules of $H_2O$ can form, consuming 300 molecules of $O_2$ and leaving 500 over.

9.16

$2 H_2 + O_2 = 2 H_2O$

$$\frac{10.0}{2.02} = 4.95 \text{ moles of } H_2$$

$$\frac{30.0}{32.0} = 0.938 \text{ moles of } O_2$$

We have more than two times the amount of $H_2$ for each unit of $O_2$. Thus $O_2$ is the limiting reactant.

We will form $(2)(0.938) = 1.88$ moles of $H_2O$

$$(1.88)(18.0) = 33.8 \text{ g of } H_2O$$

9.17

$$(10.0)(0.789) = 78.9 \text{ g}$$

The molecular weight of $C_2H_6O$ is 46.0:

$$\frac{78.9}{46.0} = 1.72 \text{ moles}$$

$$\text{Moles of } O_2 \text{ required} = (3)(1.72) = 5.16.$$

$$\text{grams of } O_2 = (32.0)(5.16) = 165.$$

handling the phenomena

1.   *Which Molecules are Polar?*
**CAUTION!** Hexane, ethyl alcohol, and benzene are flammable liquids. They should not be handled near an open flame. Smoking should not be permitted in the room while these materials are being used and a $CO_2$ fire extinguisher should be close at hand. None of the liquids should be ingested. Benzene is highly toxic and caution

should be taken not to inhale its vapors for a prolonged period of time. Dimethyl sulfoxide is rapidly absorbed through the skin and will carry other toxic materials with it. Care should be taken to avoid contact with this substance.

As explained in Section 9.3 some molecules have dipole moments. They can be represented pictorially by a little oval with one end negatively charged and the other end positively charged. If an electrically charged object is brought near such a liquid, the molecular dipoles will orient themselves in such a way as to point their end having a charge opposite to that on the object, toward the object. This will result in a net attraction between the object and the polar liquid.

Many substances can be charged simply by rubbing them on a different substance. Electrons from one substance will be transferred to the other. Thus, if a hard rubber pocket comb is rubbed with a flannel cloth, the comb will end up negatively charged. (The flannel cloth will acquire a positive charge equal in magnitude.)

Fill a burette with water. Open the stopcock so that a thin stream of water is running out of the burette as shown in Fig. 9-13. Rub a flannel cloth briskly against a hard rubber pocket comb and bring it very close to the stream of water. What do you observe? Explain your observations in terms of the discussion given above.

Hexane is a non-polar molecule. Repeat the experiment *in a fumehood*, or *other well ventilated area* replacing the stream of water with a stream of hexane. What is the effect of a charged object on a stream of non-polar molecules?

Repeat this experiment in the same well ventilated area with other liquids such as ethyl alcohol, benzene, dimethyl sulfoxide, and ethylene glycol (anti-freeze). Which ones contain polar molecules? Which of the molecules tested seems to be the most polar? Would your results have differed if you used a positively charged object?

2.   *Ionic and Covalent Substances*

**CAUTION!** Serious electrical shock and burns may result if the electrodes of the conductivity device are touched while it is plugged into an electrical outlet. ALWAYS REMOVE THE PLUG AFTER EACH TEST before washing the electrodes in distilled water.

Copper sulfate and benzene are poisonous substances and must not be ingested. Care should be taken to avoid excessive inhalation of the benzene vapors.

When an ionic substance is melted or dissolved in water, the ions become mobile—they can move around. Such a liquid or solution is a conductor of electricity.

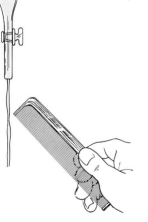

figure 9-13   A simple apparatus for the determination of polarity in molecules.

We can test for the ability to conduct electricity with the conductivity device illustrated in the accompanying diagram. It consists of a wired electrical plug, one wire of which is connected to one of the terminals of a light bulb socket; the other wire is made into an electrode. (Use a stiff wire with the insulation removed from the bottom centimeter). The other light bulb terminal is connected to a second electrode which is held parallel to and about a centimeter from the first electrode by being taped to a piece of wood.

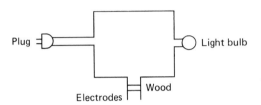

figure 9-14    A simple conductivity apparatus.

If the device is plugged into an electrical outlet, no current will flow unless the two electrodes are connected to a conducting substance.

Test each of the following to see if it conducts electricity. (The substance to be tested should be placed in a *clean* test tube and the electrodes dipped in to a distance of a few millimeters. Don't allow the electrodes to touch one another, and *don't* touch the electrodes while the plug is in. The plug must be pulled out and the electrodes washed in distilled water between each test.)

pure substances
{
Distilled water
Sugar
Salt (sodium chloride)
Copper sulfate
Benzene
}

water solutions
{
Sugar
Salt
Copper sulfate
}

Next place some sodium chloride (salt) in a crucible and heat it until it melts. Test the molten salt to see if it conducts. Repeat with sugar, taking care to heat gently so that it doesn't overheat and burn.

Explain your results. Which pure substances are ionic? Some non-ionic substances like acetic acid will break apart and form ions when dissolved in water. How could you find out if pure copper sulfate is ionic?

Assuming that salt is typical of ionic substances, and sugar of molecular substances, which type of substance melts more easily? In Chapter 10 we will make some further generalizations about different types of solids and their physical properties.

problems

1   Write an expression (similar to Equation 9-3) for the total potential energy for the hypothetical case of a molecule formed by the mutual interaction of three hydrogen atoms, i.e. three protons and three electrons.

2    Why does the potential energy rise very steeply as the intermolecular distance becomes very small for both the bonding and antibonding states, as shown in Figs. 9-1 and 9-5?

3    Draw electron dot pictures for the following molecules:

    a    H O O H          (hydrogen peroxide)

    b    F N F           (trifluoro amine)

        F

4    Which of the following pairs of elements would you expect to react to form ionic compounds?

    a    sodium (Na) and fluorine (F)

    b    calcium (Ca) and strontium (Sr)

    c    sulfur (S) and oxygen (O)

    d    silicon (Si) and carbon (C)

    e    potassium (K) and iodine (I)

5    Which of the following common gaseous molecules will have net molecular dipole moments?

    a    fluorine (diatomic $F_2$ molecule)

    b    nitrous oxide (linear NNO molecule)

    c    nitric oxide (diatomic NO molecule)

    d    nitrogen dioxide (non-linear $\underset{O\ \ O}{N}$ molecule)

    e    tetrafluoroethylene (planar $\underset{F\ \ F}{\overset{F\ \ F}{C\,C}}$ molecule)

6    Examine the following chemical equations and balance any that are unbalanced.

    a    $CH_4 + Cl_2 = CCl_4 + HCl$

    b    $NH_4NO_3 = NO_2 + 2\,H_2O$

    c    $Na + H_2O = Na^+ + OH^- + H_2$

7    What is the distinction between a molecular formula and a simplest formula?

8    When a solution of an acid such as hydrochloric acid (consisting of water molecules, $H^+$ and $Cl^-$ ions) is added to a solution of a base such as sodium hydroxide (consisting of water molecules, $Na^+$ and $OH^-$ ions) a *neutralization* reaction takes place in which the acidic $H^+$ and basic $OH^-$ ions react to form water. The $Na^+$ and $Cl^-$ ions remain in solution. Write a net ionic equation for this neutralization process.

9    Sulfur dioxide ($SO_2$) is a poisonous irritating gas which is a pollution problem in areas with industrial plants which burn fossil fuels (coal or oil) containing sulfur. How many grams of $SO_2$ will be produced when 100 kilograms of a fuel containing 0.55 percent sulfur by weight is burned?

10    What is the weight in grams of one molecule of carbon monoxide (CO)?

11    How many molecules of water are there in a single drop? (Drops vary in size. Consider

an average sized drop such as that formed by a medicine dropper which has a volume of about $0.08$ cm$^3$). Water has a density of $1.0$ g/cm$^3$.

12    A certain copper ore is found to contain 18 percent cupric oxide (CuO) by weight. How many moles of CuO are contained in 1.0 kilogram of this ore? How many kilograms of ore would be needed to extract 1.0 kilogram of copper by means of the reaction

$$CuO + C = Cu + CO$$

and how many grams of carbon monoxide would be produced?

# the states of matter

In Chapter 1 it was noted that under the conditions of temperature and pressure existing on the surface of the earth, most matter can be classified as being in either the gaseous, liquid, or solid state. It was pointed out that, as is the case with most classification schemes, we occasionally encounter examples that don't fit neatly into the system. For the vast majority of common materials, a clearcut designation of state *is* possible, and it is most useful to examine the general properties that distinguish the three common forms of matter from one another.

## 10.1  gases

In the gaseous state, matter is in its least structured, most diffuse form. A gas will expand indefinitely to fill ever larger containers of any conceivable shape. We now know that gases consist of widely separated molecules in constant chaotic motion. This knowledge has resulted from painstaking experimentation and theorizing over a period of several centuries.

**Boyle's Law (Pressure-Volume Relationship).**  The earliest description of an important general property of gases is attributed to the seventeenth century English scientist-alchemist, Robert Boyle. Boyle made measurements on the relationship between pressure and volume of several common gases (or "airs" as they were then called). Pressure is a measure of a force exerted per unit area over a surface. For example, the pressure due to the atmosphere measured at sea level on an average day is about 101,325 newtons per square meter. This is equal to the pressure exerted by a column of mercury 76.0 cm high which defines the unit of pressure called the *atmosphere* (atm.). Another frequently used unit of pressure is the *torr*, which is equal to the pressure exerted by a column of mercury one millimeter in height. Thus one torr equals 1/760 atm. A confined gas will exert an equal pressure over the entire surface of its container.

Boyle concluded that the pressure exerted by a given quantity of any confined gas is inversely proportional to the volume it occupies. This observation (commonly referred to as Boyle's Law) can be represented by the equation:

$$P = \frac{a}{V} \qquad \text{(or PV} = a)  \qquad (10\text{-}1)$$

in which $P$, $V$, and $a$ represent respectively the pressure, volume, and a constant. The value of $a$ depends on the amount of gas and the temperature, as well as on the units used to measure the pressure and the volume. This relationship is illustrated graphically in Fig. 10-1.

**Charles' Law (Volume-Temperature Relationship).** Not until the end of the eighteenth century was the quantitative relationship between temperature and gas volume investigated. The French scientist, Jacques Charles, performed some rather crude experiments in 1787 on the basis of which he proposed that if the pressure is held constant, the volume of a gas varies linearly with the temperature. Much more careful work performed a few years later by another Frenchman, Joseph Louis Gay-Lussac, confirmed Charles' proposal. The volume-temperature relationship (Charles' Law) is shown graphically in Fig. 7-6. If the Kelvin temperature scale is used, where 0° equals absolute zero, (see Chapter 7), Charles' Law takes the simple mathematical form:

$$V = bT \qquad (10\text{-}2)$$

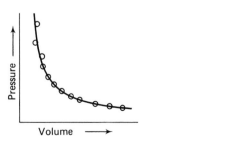
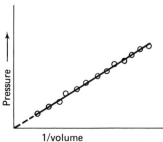

figure 10-1   The pressure-volume relationship for a gas.

in which $b$ is a constant. The value of $b$ is dependent on the amount of gas and the pressure, as well as on the units used to measure the volume and the temperature.

*Question 10.1*  What is the equation representing Charles' Law if the temperature scale used is Celsius or Fahrenheit?

**Universal Gas Law.**  The interrelationships among temperature, pressure, and volume contained in Boyle's and Charles' Laws for any sample of gas can be written in the form of a single combined equation,

$$\frac{PV}{T} = c \tag{10-3}$$

The constant, $c$, is dependent on the amount and kind of gas and it is probably not surprising to learn that for a given gas the value of $c$ is directly proportional to the mass of the gas sample. A more startling observation is that if we express the amount of gas in moles (designated by the symbol $n$) we can convert Equation 10-3 into the equation:

$$\frac{PV}{T} = nR \tag{10-4}$$

which has a *unique* constant, $R$, applicable to all gases! $R$ is called the universal gas constant. If volume is measured in liters, pressure in atmospheres, and temperature in degrees Kelvin, the numerical value of $R$ is found to be 0.0821 (atmosphere · liters)/(degree · mole).

*Question 10.2*  Show that the pressure-volume relationship of Equation 10-1 and the volume-temperature relationship of Equation 10-2 are embodied in Equations 10-3 and 10-4.

*Question 10.3*  What will be the volume occupied by 32.0 grams of oxygen gas at 0°C and a pressure of 1.0 atm.?

**Avogadro's Hypothesis (Equal Volumes of Gases Contain Equal Numbers of Molecules).**  Note that Equation 10-4 demands that equal volumes of two different gases at the same temperature and pressure must contain the same number of moles, and, therefore, the same number of molecules as illustrated by Fig. 10-2. As early as 1811 the Italian physicist Amedeo Avogadro hypothe-

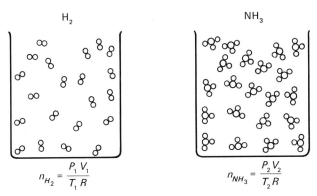

figure 10-2   If $P_1 = P_2$, $T_1 = T_2$, and $V_1 = V_2$, then
$n_{H_2} = n_{NH_3}$. Since the number of molecules is proportional to $n$, the two equal volumes of the different gases
must contain the same number of molecules.

sized that this was true. Gay-Lussac's remarkable discovery that the combining
*volumes* of gases at any fixed temperature and pressure could be expressed
in simple whole number ratios led Avogadro to his hypothesis. For instance,
it is found that 1.0 liter of ammonia gas, $NH_3$, will combine chemically with
1.0 liter of hydrogen chloride gas HCl, and that 1.0 liter of oxygen gas, $O_2$,
will combine with 2.0 liters of hydrogen gas, $H_2$. In Chapter 8 it was shown
that, in general, the quantities of two substances that combine chemically will
be such that the ratio between the number of *moles*—and therefore *molecules*—in the two samples is equal to the ratio of the small integers preceding
the formulas of the substances in the chemical equation for the reaction.
Avogadro knew that no simple numerical relationship existed between the
combining *volumes* of solids and liquids. He correctly deduced that this unique
volume relationship of gases could only be reconciled with the atomic theory
if it was assumed that one mole of *any* gas at a fixed temperature and pressure
would always occupy the *same volume*. (Unfortunately for Avogadro, and for
the understanding of both atomicity and gaseous structure, the validity of his
argument was not widely accepted until almost 50 years later!)

*Question 10.4*   If 1.0 liter of nitrogen gas, $N_2$, is found to combine with 2.0 liters of
oxygen gas, $O_2$, to form 1.0 liter of a product gas, what is the formula of
this gas?

**Problems Involving Gas Law Calculations.**   Students of chemistry and
physics are frequently plagued by one of a wide variety of examination or
homework problems calling for the calculation of a pressure, volume, temperature, or amount of a gas. A careful reading of these problems will invariably

reveal that three of the four variables, $P$, $V$, $T$, and $n$ are given either directly or indirectly. Equation 10-4 can then be employed to determine the value of the fourth variable. To make the student's life a bit more miserable, devious problem writers frequently combine "gas law problems" with "chemical" problems such as those exemplified in Chapter 9. Examples of two such problems and their solutions follow.

*Example 10.1* Solid calcium hydride, $CaH_2$, reacts with water, $H_2O$, to form solid calcium hydroxide, $Ca(OH)_2$, and hydrogen gas, $H_2$. How many grams of water would have to be added to excess calcium hydride to produce enough hydrogen to fill a balloon to a volume of 7.5 liters at a pressure of 1.2 atm. and a temperature of 25°C?

*Solution* We are given the values of $P$(1.2 atm.), $V$(7.5 liters), and $T$(25°C = 298°K). We can therefore determine the number of moles, $n$, of hydrogen.

$$n = \frac{PV}{RT} = \frac{(1.2)(7.5)}{(.082)(298)} = 0.037 \text{ moles}$$

The balanced equation for the chemical reaction is:

$$CaH_2 + 2H_2O = Ca(OH)_2 + 2H_2$$

The coefficients of $H_2O$ and $H_2$ in the equation are both 2. This indicates that the number of moles of water required is exactly equal to the number of moles of hydrogen gas produced. Thus 0.037 moles of $H_2O$ is required.

amount (in grams) = (number of moles)(molecular weight in grams/mole)

grams $H_2O$ = (0.037)(18) = 0.67

*Example 10.2* When potassium chlorate, $KClO_3$, is heated, it is converted to potassium chloride, $KCl$, and oxygen gas. The oxygen produced from 15.0 grams of potassium chlorate is found to occupy a volume of 5.00 liters at a pressure of 1.00 atm. What is the temperature of the oxygen gas?

*Solution* The balanced equation is:

$$2KClO_3 = 2KCl + 3O_2$$

The $P$ and $V$ of the $O_2$ are given. If we knew $n$ we could calculate $T$. The formula weight of $KClO_3$ = 123. Moles of $KClO_3$ = 15.0/123 = 0.122. From the equation we see that $1\frac{1}{2}$ moles of $O_2$ are produced from each mole of $KClO_3$. Therefore, $n = \frac{3}{2}(0.122) = 0.183$ moles of $O_2$.

$$T = \frac{PV}{nR} = \frac{(1.00)(5.00)}{(0.183)(.0821)} = 333°K = 60°C$$

**Cannizzaro's Method for Determining Atomic Weights.**   We can also use Equation 10-4 to determine the molecular weight of a gaseous compound from a measurement of its density (mass per unit volume) at any given temperature and pressure. In order to see that this is so, we will first rearrange Equation 10-4 to the form:

$$\frac{n}{V} = \frac{P}{RT} \tag{10-5}$$

Next we can multiply both sides of the equation by the molecular weight, $W_m$, of the gas in question:

$$\frac{W_m(n)}{V} = \frac{PW_m}{RT} \tag{10-6}$$

Since the molecular weight multiplied by the number of moles gives the mass of the sample, we have on the left side of the equation, mass divided by volume, which is the density, $\delta$.

$$\delta = \frac{PW_m}{RT} \tag{10-7}$$

A final rearrangement yields the equation for the molecular weight:

$$PW_m = \delta RT$$

$$W_m = \frac{\delta RT}{P} \tag{10-8}$$

An experimental method for measuring molecular weights based on Equation 10-8 is described in Section 1 of Handling the phenomena at the end of this chapter. It was the use of this relationship that led Stanislao Cannizzaro, a student of Avogadro, to a useful method for establishing both atomic weights and molecular formulas. If you know the molecular weight of a substance, and its percentage composition, you can determine a number for each element in the molecule which is equal to its atomic weight multiplied by the number of atoms of that element in the compound. For example, ethyl alcohol is composed of 13.1 percent hydrogen, 34.8 percent oxygen, and 52.1 percent carbon, and has the formula $C_2H_6O$ and a molecular weight of 46.1. This means that 13.1 percent of 46.1 or 6.05 is the part of the molecular weight contributed by the hydrogen atoms in the molecule. Thus, 6.05 must equal the number of hydrogen atoms in ethyl alcohol times the atomic weight of hydrogen ($6 \times 1.008 = 6.05$). Similarly, 34.8 percent of 46.1, or 16.0, equals the number of oxygen atoms in ethyl alcohol times the atomic weight of oxygen ($1 \times 16.0 = 16.0$) and 52.1 percent of 46.1 or 24.0 equals the number of atoms of carbon in ethyl alcohol, times the atomic weight of carbon $(2 \times 12.0 = 24.0)$. These relationships are illustrated in Fig. 10-3.)

Suppose you don't know the atomic weight of chlorine. You do know the percentage (by weight) of chlorine in five different volatile compounds, but have no idea what the molecular formulas of these compounds are. (This is the sort of situation that existed in Cannizzaro's day.) The densities of gaseous samples of the five compounds can be determined at a known $T$ and $P$, and then Equation 10-8 can be used to find the corresponding molecular weights. By multiplying the molecular weight by the percentage of chlorine in each compound five numbers will be obtained. Each of these numbers will equal a small whole number (the number of atoms of chlorine in the molecule), times the atomic weight of chlorine. If you included a molecule containing only one atom of chlorine the number you obtain for this compound will be the atomic weight of chlorine and it will be a common divisor of the other numbers. Even if you don't include a compound with only one atom of chlorine it is likely that the largest common divisor will be the atomic weight of chlorine. This method is illustrated in Table 10-1. It is the procedure suggested by Cannizzaro to determine the atomic weights of several of the lighter elements.

*Question 10.5*   Verify the numbers that appear in the density column in Table 10-1. Use the given values in the column of molecular weights.

*Question 10.6*   Suppose that molecules always contained at least two atoms of a particular element. Is Cannizzaro's method of atomic weight determination capable of detecting this?

**Additional Properties of the Gaseous State.**   Before proceeding to a theoretical explanation of the behavior of gases it will be useful to examine

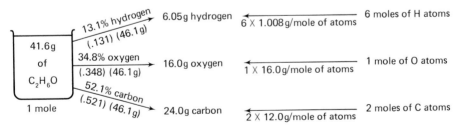

figure 10-3   Since moles are proportional to molecules or atoms, this illustration shows that the percentage by weight of an element in a molecule multiplied by the molecular weight of the molecule must equal the number of atoms of that element in the molecule multiplied by the atomic weight of the element.

table 10.1    determination of the atomic weight of chlorine by cannizzaro's method.

| Chlorine Containing Compound | % Cl by Weight | Density (grams/liter) at P = 1.00 atm t = 100°C | Molecular Weight | $\dfrac{(\%Cl)(W_M)}{100}$ |
|---|---|---|---|---|
| A | 100 | 2.32 | 71.0 | 71.0° |
| B | 89.2 | 3.90 | 119 | 106 ° |
| C | 92.2 | 5.04 | 154 | 142 ° |
| D | 83.5 | 2.78 | 85.0 | 71.0° |
| E | 90.8 | 3.84 | 117 | 106 ° |

° Largest common divisor of numbers in last column is 35.5 which is the atomic weight of Cl.

some other empirical observations that provided additional clues to theorists.

The first of these is the observation, originally due to John Dalton, that the total pressure exerted by a mixture of gases in a container is equal to the sum of their *partial pressures*. The partial pressure of a gas is that pressure which each of the component gases would exert if it were the only gas present in the container. Another way of visualizing this observation is to consider a collection of containers, all of the same volume, with each one containing a different gas at the different pressures $P_1$, $P_2$, and $P_3$. If all of the gases are transferred to one of the containers, the total pressure in it would be $P_1 + P_2 + P_3$, as shown in Fig. 10-4. (*Note:* This observation is not trivial. Physical properties are *not* necessarily additive. For example, the volume of a liquid solution is often considerably less than the sum of the separate volumes of its constituent parts. Anyone who has added sugar to a cup of coffee filled to the brim has been in a position to observe this phenomenon.) Equation 10-4 is applicable to gas mixtures, and can relate the total pressure to the total number of moles of gas present, or the partial pressure of any one gas to the number of moles of that gas.

*Example 10.3* A 1.00 liter flask initially contains air at a temperature of 27°C and a pressure of 1.000 atm. The air is pumped out of this flask and forced through a tube containing red hot copper granules which react with and remove all of the oxygen. The deoxygenated air is returned to the 1.00 liter bulb and allowed to cool to 27°C. The final pressure is found to be 0.790 atm. How many moles of oxygen were in the original air sample?

*Solution* From Dalton's Law we know that the oxygen initially in the bulb must have exerted a partial pressure, $P_{O_2}$, equal to the difference between the total pressure, $P_T$, and the pressure due to the gases in the deoxygenated gas, $P_d$.

$$P_{O_2} = P_T - P_d$$
$$P_{O_2} = 1.000 - 0.790 = 0.210$$

Using Equation 10-4 we find:

$$P_{O_2} = n_{O_2} \frac{RT}{V}$$

or

$$n_{O_2} = \frac{P_{O_2}V}{PT} = \frac{(0.210)(1.00)}{(0.0821)(300)} = 0.00853 \text{ moles}$$

Anyone who has driven on a country road and detected the recent presence of a skunk, or who has gone through the halls of an apartment building at dinner time and has become aware of what each family is eating, is familiar with the fact that gases diffuse. The Scottish chemist-mathematician, Thomas Graham, is responsible for empirically discovering that the rates of diffusion of gases at a given temperature are inversely proportional to the square roots of their molecular weights. For any two gases Graham's Law states that the rates of diffusion ($r_1$ and $r_2$) will be related to their molecular weights ($W_{M_1}$ and $W_{M_2}$) by the equation:

$$\frac{r_1}{r_2} = \sqrt{\frac{W_{M_2}}{W_{M_1}}} \tag{10-9}$$

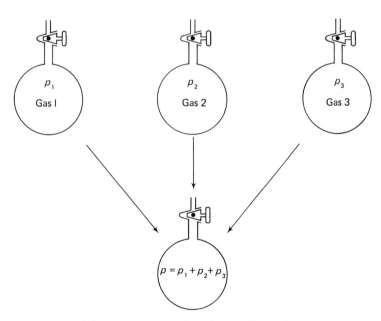

figure 10-4   If the gases contained in three different flasks of equal volume are transferred to a fourth flask of the same volume, the pressure in this flask will be the sum of the pressures that the gases exerted in their individual flasks before the transfer.

*Question 10.7*  Describe a demonstration that could be performed in a lecture hall to show the relationship between gas diffusion rate and molecular weight in a semiquantitative manner.

Another interesting phenomenon can be observed by using a low power microscope to view the little dance performed by tiny smoke particles (from a match, or cigarette) suspended in calm air made visible by a sharply defined, bright beam of light. The English botanist, Robert Brown, studied the analogous erratic motions of tiny grains of pollen suspended in water. He first thought that the movement of the pollen grains could be attributed in some way to the fact that pollen is derived from living matter. But later he found that any material, when sufficiently finely divided, would jump around when suspended in liquids or gases. A theory that purports to explain the general behavior of gases should also be able to explain this so-called *Brownian motion*. An example of Brownian motion is diagrammed in Fig. 10-5.

**The Kinetic Theory of Gases.**   We have now considered a large number of general properties of gases. The search for a theoretical model to provide a unified understanding of gas-phase phenomena was intertwined with the evolution of the explanation of heat phenomena discussed in Chapter 7. Many of the same scientific workers were involved. Two basic models were proposed. One was a *static model* in which a gas was considered to be either a collection of stationary molecules that were either in contact with each other and capable of expanding and contracting, or a collection of molecules that could repel

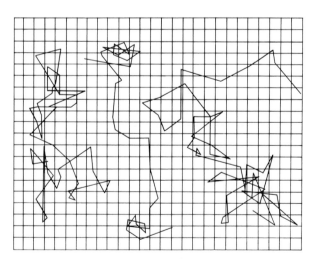

figure 10-5  Brownian movement. Highly magnified plot of the position of three minute particles suspended in water, observed at 30 second intervals. (After J. Perrin)

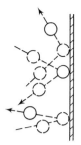

figure 10-6   Bernoulli ascribed gas pressure to the incessant collisions of minute spherical gas particles with the walls of the container.

each other at a distance. The second was a *kinetic model*, in which the
molecules were separated and in constant violent motion. Boyle described,
for both models, the qualitative features necessary to account for his
pressure-volume observations. Newton (although he appeared to prefer the
kinetic model) showed that the static approach, with molecules that repelled
each other with a force that was inversely proportional to their separations,
could account for the inverse pressure-volume relationship.

The Swiss physicist, Daniel Bernoulli, published a treatise in 1738 based
on the kinetic model which had many of the features of the presently accepted
theory of gases. He envisioned that the pressure shown by a gas arose from
the incessant collisions of minute spherical gas particles with any surface that
was in contact with the gas (as shown in Fig. 10-6.) He ascribed the rise in
pressure with decreasing volume to the increase in frequency of these collisions
as the average intermolecular separation decreases. Bernoulli was able to
deduce Boyle's Law mathematically from his assumptions about the relationship between gas pressure and molecular motion.

Despite this apparent success, Bernoulli's notions were not in accord with
those of his contemporary scientists. There were insufficient data available
to test other consequences of this quantitative kinetic model. For the next
hundred years the debate concerning alternative models of gas structure
continued with considerable sentiment, expressed by such luminaries as Dalton,
for a static model.

With the clarification of the nature of heat (described previously in Chapter
7), and the establishment of the equivalence between heat and mechanical
energy, scientists showed a renewed interest in the kinetic approach to the
explanation of gas behavior. Indeed, during the middle of the nineteenth
century, it was Joule himself who was the principal figure in resurrecting and
extending Bernoulli's theory. Joule's work provided the background for the
publication of a sophisticated and far-reaching kinetic theory of gas behavior
by the German physicist, Rudolph Clausius. With some further elaboration
through the work of such physicists as James Clerk Maxwell (British), Ludwig
Boltzmann (Austrian), and J. Willard Gibbs (American), Clausius' theory
provides the basis for our present understanding of the general properties of
gases.

The basic postulates of the kinetic model of gases are as follows:

a    *All Gases Consist of Molecules*

This postulate was originally based primarily on the chemical evidence provided by Dalton, Gay-Lussac, Avogadro, and others.

b    *The Size of Gas Molecules is Negligible When Compared with the Distance Between Them*

It is observed that, when vaporized, a drop of liquid will produce a gas that occupies a volume about 1000 times as great (at atmospheric pressure). If the pressure is reduced the volume can be increased ad infinitum. Unless one makes the awkward assumption that the gas molecules themselves can expand in this limitless manner, it follows that most of a gas must be empty space.

c    *The Molecules are in Constant Random Straight Line Motion*

This property is suggested by the observed diffusion of gases and their ability to expand indefinitely to fill the available volume. This direction of motion will change following each collision so that an individual molecule is thought of as describing a zig-zag path through space as shown in Fig. 10-7.

d    *All Intermolecular Forces are Negligible Except During the Instant of Collision*

This assumption means that the velocity of a gas molecule will remain constant between collisions. Its validity is born out both by the success of the theory and by calculations of the magnitude of intermolecular forces (both electrical and gravitational) based on our present knowledge of molecular structure.

e    *On the Average, Collisions Between Molecules and With the Walls of the Container are Perfectly Elastic*

As defined in Chapter 2, an elastic collision is one in which the total kinetic energy of the interacting bodies is conserved. The need for this postulate becomes apparent if we consider what would happen if molecular collisions resulted in even a gradual loss of kinetic energy. Under these circumstances the average velocity of the molecules in a gas would continually decrease. All of the more sophisticated kinetic models retained Bernoulli's understanding of gas pressure as resulting from molecular collisions. If the average molecular velocity decreased, then the pressure of a gas in a thermally insulated closed container should drop. This is not observed, and so we are led to the necessity of this fifth postulate. (For any given collision, some of the kinetic energy might produce excited internal

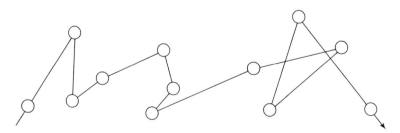

figure 10-7    The zig-zag path taken by a gas molecule. Each change of direction is due to a collision with another molecule.

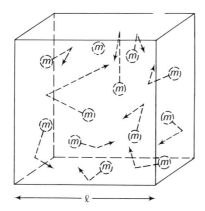

figure 10-8   The model discussed in the text consists of $N$ molecules, each of mass, $m$, and moving with speed, $\bar{v}$, in random directions, in a cubical container with edges of length $\ell$.

motion of the molecule. On the next collision, the excitation energy of this molecule might be given back to the kinetic energy of one of the colliding molecules. On the average, no kinetic energy is lost.)

*Question 10.8*   In the earlier chapters we found that collisions between even the hardest macroscopic objects are only approximately elastic. (A steel ball bouncing on a steel platform gradually loses kinetic energy and comes to rest.) How then can we even entertain a postulate that requires perfectly elastic collisions between molecules?

   A completely rigorous and general analysis of the behavior of a gas based on the five postulates requires the use of sophisticated mathematical tools. It is possible, however, to demonstrate, using simple algebra and some approximations that can be shown to be completely justified, that in the special case of a cubical container the model leads to an equation for the pressure-volume behavior of the form given by Equation 10-1, and that with one additional postulate Equation 10-4 can be obtained. This derivation is similar to the one developed by Joule.
   Consider a gas consisting of $N$ identical molecules, each of mass $M$, moving with an average speed $\bar{v}$, in random directions in a cubical container, with edges of length, $\ell$. This situation is illustrated in Fig. 10-8. These molecules will strike the six walls of the container at every conceivable angle. To simplify our treatment, let us replace this completely random motion by a situation in which the molecules are divided into three equal groups, $\frac{1}{3}N$, each of which is restricted to a back and forth motion perpendicular to one of the three sets of opposite walls, as shown in Fig. 10-9.
   Let us now zero in on one molecule which, in our simplified model, will be moving back and forth between two opposite faces of the container. If this molecule is traveling with speed $\bar{v}$ and undergoes an elastic collision with

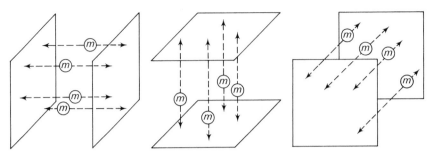

figure 10-9    Simplified model, $\frac{1}{3} N$ molecules moving back and forth perpendicular to each pair of parallel walls in the cube.

the wall, it will simply change direction without altering its speed. If it has momentum $+M\bar{v}$ before the collision (moving towards the right), it will have momentum $-M\bar{v}$ after the collision, as shown in Fig. 10-10). For each collision,

$$\Delta \text{ momentum} = \quad -M\bar{v} \quad - \quad (+M\bar{v})$$

| (Change in Momentum) | (Final Momentum) | (Initial Momentum) | (10-10) |

$$= \quad -2M\bar{v}$$

Back in Chapter 5 we found that when an object undergoes a change in momentum we can determine the force it exerts during that change by using the formula,

$$F = -\frac{\Delta \text{ momentum}}{\Delta t} \tag{10-11}$$

in which $\Delta t$ is the time interval involved in the momentum change. The force exerted on the right-hand wall, $F_w$, will be equal to:

$$F_w = -\left(\frac{-2M\bar{v}}{\Delta t}\right) = \frac{2M\bar{v}}{\Delta t} \tag{10-12}$$

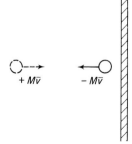

figure 10-10    A molecule having momentum equal to $+M\bar{v}$ before collision with a wall will have momentum equal to $-M\bar{v}$ afterwards.

If we can determine how many molecules strike the wall during a finite time interval, we can use Equation 10-12 to determine the average force on the wall. Suppose all of the molecules moving perpendicular to one pair of opposite walls are allowed to travel a distance equal to $2\ell$. In so doing they will have collided with each wall exactly once no matter what position they start from (see Fig. 10-11). If the average molecular speed is $\bar{v}$ the time, $\Delta t$, involved will be

$$\Delta t = \frac{\text{distance}}{\text{speed}} = \frac{2\ell}{\bar{v}} \qquad (10\text{-}13)$$

During this time interval the number of collisions with either wall will be equal to the total number of molecules confined by our model to moving between a particular pair of walls. This number is $\frac{1}{3}$ of the molecules present in the container, or $N/3$. Since each collision results in a momentum change of magnitude $2M\bar{v}$, the total momentum change will be:

$$\Delta \text{ momentum} = \left(\frac{N}{3}\right)(2M\bar{v}) \qquad (10\text{-}14)$$

The average force on any one wall, $F_w$, will be:

$$F_w = \frac{\Delta \text{ momentum}}{\Delta t} = \frac{\left(\dfrac{N}{3}\right)(2M\bar{v})}{\left(\dfrac{2\ell}{\bar{v}}\right)} \qquad (10\text{-}15)$$

$$= \frac{N}{3}(2M\bar{v})\frac{\bar{v}}{2\ell}$$

$$= \frac{2NM\bar{v}^2}{6\ell}$$

$$= \frac{NM\bar{v}^2}{3\ell}$$

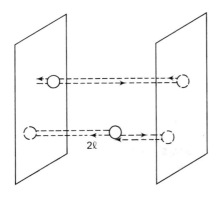

figure 10-11   No matter what position a molecule starts at, it will have collided with each opposite wall exactly once after traveling a distance $2\ell$.

To compute the pressure, $P$, recall that pressure is force divided by area, and that the area of the square wall, $A_w$, of the cubical container is $\ell^2$. Thus:

$$P = \frac{F_w}{A_w} = \frac{\left(\frac{NM\bar{v}^2}{3\ell}\right)}{\ell^2} = \frac{NM\bar{v}^2}{3\ell^3} \tag{10-16}$$

Note that the denominator of this last equation contains the term $\ell^3$. This is equal to the volume, $V$, of the container. By making this substitution we arrive at the equation:

$$P = \frac{NM\bar{v}^2}{3V} \tag{10-17}$$

in which the pressure and volume are inversely related, as found by Boyle! Let's play around a bit more with the equations. If we rewrite Equation 10-17 in the form:

$$P = \left(\frac{2N}{3V}\right)\tfrac{1}{2}M\bar{v}^2 \tag{10-18}$$

we can recognize that the quantity $\tfrac{1}{2}M\bar{v}^2$ is the average kinetic energy of the gas molecules. Thus we find that the pressure is proportional to the average molecular kinetic energy. (The quantity $2N/3V$ is the proportionality constant.) In Chapter 7 we developed a kinetic model of heat in which we postulated that the average kinetic energy of the molecules of any substance is proportional to a variable which we call the absolute temperature, $T$. Since a proportionality constant is simply a number, we can choose any symbol we want to use for it. Other work has shown that there are good reasons (related to the nature of the space that the molecules are considered as being in) for choosing $\tfrac{3}{2}k$ as the proportionality constant between the molecular kinetic energy and the absolute temperature.

$$\tfrac{1}{2}M\bar{v}^2 = \tfrac{3}{2}kT \tag{10-19}$$

Invoking this final postulate, and making the appropriate substitution in Equation 10-18, we arrive at:

$$P = \left(\frac{2N}{3V}\right)\left(\frac{3}{2}kT\right)$$

$$P = \frac{6N}{6V}kT$$

$$PV = NkT \tag{10-20}$$

which we can rearrange to:

$$\frac{PV}{T} = kN \tag{10-21}$$

All that remains is to observe that we now have an equation which relates
$P$, $V$, and $T$ to the *number* of molecules present. As we discovered in Chapter
9, this number $N$ is very large for laboratory sized samples, and it is convenient
to convert from molecules ($N$) to moles ($n$). By recalling that
$n = N/6.02 \times 10^{23}$ we can write:

$$\frac{PV}{T} = k(6.02 \times 10^{23}) \left( \frac{N}{6.02 \times 10^{23}} \right), \tag{10-22}$$

or

$$\frac{PV}{T} = k(6.02 \times 10^{23})n \tag{10-23}$$

If we now identify $k(6.02 \times 10^{23})$ as the constant, $R$, we have exactly reproduced the empirically observed expression given by Equation 10-4 by applying
the postulates of the kinetic model and a few physical relationships to a gas
in a cubical container, making a few simplifying assumptions, and using some
simple algebra! The fact that we have considered only cubical containers
should not trouble anyone. Indeed, any differently shaped container can be
subdivided into infinitesimal little cubes.

The success of a more general treatment (without our assumptions but using
the same postulates) in producing the same result, Equation 10-23, was
instrumental in gaining widespread acceptance of the kinetic model. Further
confidence in this theory resulted from the ease with which it could be used
to explain other gaseous phenomena. For example, we can explain Graham's
Law, Equation 10-9, as follows:

It seems reasonable to expect that the rate at which a gas diffuses will
be proportional to the average speed with which its molecules are moving.
A rigorous treatment of the dynamics of a gas in which the moving, colliding
particles are describing a so-called *random walk,* such as we assumed for our
kinetic model (Fig. 10-7), verifies this intuitive expectation. Therefore, we can
substitute the ratio of average molecular speeds, $\bar{v}_1/\bar{v}_2$, for the ratio of
diffusion rates, $r_1/r_2$ in Graham's Law, Equation 10-9. Also note that since
molecular weights are proportional to the average masses of the individual
molecules, the ratio of molecular weights, $W_{M_2}/W_{M_1}$ will also be the ratio
of the masses of the molecules.

$$\frac{\bar{v}_1}{\bar{v}_2} = \sqrt{\frac{W_{M_2}}{W_{M_1}}} \tag{10-24}$$

Our two diffusing gases will both be at the same temperature and using
Equation 10-19 we see that:

$$\frac{1}{2} W_{M_1} \bar{v}_1^2 = \frac{1}{2} W_{M_2} \bar{v}_2^2 \tag{10-25}$$

Dividing both sides of this equation by $\frac{1}{2}M_1\bar{v}_2^2$ we get:

$$\frac{\bar{v}_1^2}{\bar{v}_2^2} = \frac{W_{M_2}}{W_{M_1}} \tag{10-26}$$

If we take the square root of both sides of this equation we obtain Equation 10-24! Thus we have deduced the empirical result (Equation 10-9 or Equation 10-24) from our theoretical model (Equation 10-19).

*Question 10.9*    Demonstrate that Dalton's law of partial pressures and Brownian motion are both expected consequences of the kinetic theory of gases.

**Real Gases—Deviations from Ideal Behavior.**    Now that we have gone to the trouble of developing a fairly simple model that leads exactly to Equation 10-4 it is time to muddy the clear water by pointing out that real gases "obey" this equation only over a limited range. A graph of $PV$ vs $P$, at constant $T$, should be a straight horizontal line if Equation 10-4 is valid. In Fig. 10-12a a graph for such an "ideal gas" is shown. In contrast Figs. 10-12b, c, and d illustrate deviations for real gases from this ideal behavior. We see that $N_2$ at 20°C adheres closely to the ideal line until about $P = 50$ atm., and then deviates only slightly until $P = 200$ atm. The curve for this same gas at −70°C shows much greater deviation from the "ideal" line as does the curve for $CO_2$ at 40°C.

In general we find that all real gases obey Equation 10-4 at very low pressures. Negative deviations (dips below the ideal line) become less pronounced as the temperature is raised far above the point at which the gas can be liquified. ($N_2$ cannot be liquified at temperatures greater than −147°C, whereas $CO_2$ can be liquified at 31°C.) Positive deviations do not become very pronounced for real gases until the pressure is raised to over 200 atm.

*Question 10.10*    Which of the following gases would you expect to show large negative deviations from the "ideal gas" curve at the temperature indicated: water at 110°C, oxygen at 50°C, helium at −50°C?

The source of the negative deviations from the $PV$ values predicted by Equation 10-4 lies in postulate (d) of our kinetic model. The ideal gas molecules were assumed to exert no intermolecular forces. Fig. 10-13 shows in a qualitative way how the force between real molecules varies with distance. We see that molecules experience an attractive force as they approach each other. This interaction, called *van der Waals' forces* after the Dutch physicist who investigated it, has its origin in the electromagnetic interaction, and varies

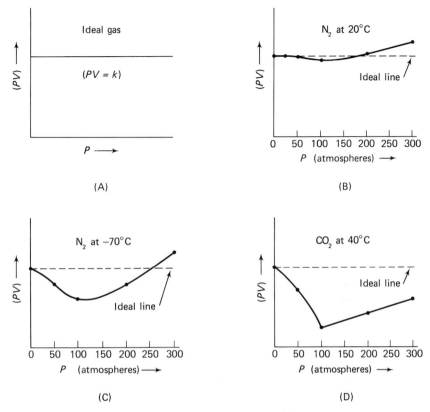

figure 10-12    Deviations of real gases from ideal gas behavior. (From *Principles of Physical Science*, by F. T. Bonner and M. Phillips. 1957. Addison-Wesley.)

greatly from molecule to molecule. In general the magnitude of the van der Waals' attraction is roughly proportional to the number of electrons in the molecules of the gas, but it is also dependent on specific properties related to molecular structure and geometry. (Molecules also exert gravitational forces on one another, but the van der Waals' force is orders of magnitude greater.) The fact that real gas molecules attract each other tends to make it easier to compress a real gas than an ideal one, and therefore results in a decrease in the *PV* product. We see from Fig. 10-13 that the attractive force is very small at the average intermolecular distance between molecules in the gaseous state. As the molecules approach each other this force increases rapidly. At low temperatures, and therefore low kinetic energies, the attractive force between molecules will be sufficiently great to slow them up. They will thus spend a relatively greater proportion of their time in close proximity where the attraction is large, and thus large negative deviations from ideal behavior will be observed. At high temperatures the kinetic energy of the molecules

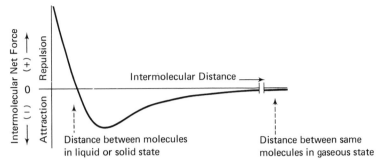

figure 10-13   How the intermolecular force varies with the distance between molecules. (From *Foundations of Modern Physical Science* by G. Holton and D. H. D. Roller. 1958. Addison-Wesley.)

will be large relative to the attractive force, and the motion of the molecules will be almost unaffected by the attractive force, resulting in nearly ideal behavior.

The positive deviations are even easier to explain. Postulate (b) is based on our experience that gases at *atmospheric* pressure consist of mostly empty space, and therefore the actual volume occupied by the gas molecules themselves can be neglected. As we increase the pressure on the gas we are forcing the molecules into a smaller and smaller volume. At sufficiently high pressures the total volume occupied by all of the individual gas molecules will no longer be insignificant, and it will become more difficult to compress the real gas than an ideal gas. At very high pressures we therefore observe a PV product greater than that predicted by Equation 10-4.

Various modifications of Equation 10-4 have been proposed that result in an accurate representation of the pressure-volume-temperature behavior of real gases over a wider range of conditions. One such expression, which was proposed by van der Waals in 1873, is:

$$\left(P + \frac{n^2a}{V^2}\right)(V - nb) = nRT \qquad (10\text{-}27)$$

where $a$ and $b$ are constants that assume different values for each gas. (Note: Do not confuse the van der Waals constants, $a$ and $b$, with the constants used previously in equations 10-1 and 10-2.)

*Question 10.11*    The experimentally determined values of $a$ and $b$ for nitrogen gas are 1.390 liter$^2 \cdot$ atm/mole$^2$ and 0.03913 liter/mole respectively. Compare the values of PV calculated from Equation 10-4 with those calculated from Equation 10-27 for a 1.00 mole sample of nitrogen gas confined to a 1.00 liter container at (1) 50°C and (2) −100°C.

# 10.2   liquids

Introductory physical science texts frequently contain rather detailed discussions of solids and gases but devote little, if any, attention to the liquid state. Why is this? Is the liquid state less important? Indeed, it is true that almost all of the elements in the periodic table are either solids or gases under ordinary conditions—and this applies as well to the compounds formed from the elements.

*Question 10.12*   Which of the elements in the periodic table exist as liquids at room temperature and one atmospheric pressure? See how many pure liquid compounds, you can name that you have actually encountered outside a science laboratory.

It is also true, however, that the vast majority of biologically important processes involve liquids, that seven tenths of the surface of the earth is covered by liquid, and that liquids have an enormous effect on the geology of the land areas as well. It is therefore not true that liquids receive less attention because they are less important.

Gases, as we have seen, are assumed (on first approximation) to have no significant internal structure. Our ideal model for their behavior is based on random thermal motions. As we shall see, the solid state is also describable in terms of an ideal model. The perfect crystal has a highly regular structure on which the thermal motions of atoms and molecules can be considered to have only a slight perturbing effect. The liquid state is intermediate, in that both structure and molecular motion are of comparable importance. No simple ideal liquid model has a yet emerged that would serve the same unifying function as the ideal gas model, or the perfect crystal model. It is this relative complexity of the liquid state that leads to its neglect in introductory texts.

**Some General Properties of Liquids.**   Unlike gases, liquids will not expand to completely fill any container in which they are placed. A given quantity of liquid has its own definite volume (and density) at any given temperature and pressure. A liquid suspended in space with no external forces acting on it would adopt a spherical shape. When placed in a container on the surface of the earth gravitational forces result in its conforming to the shape of the container as illustrated in Fig. 10-14.

The densities of liquids are much greater than the densities of corresponding gases under the same conditions of temperature and pressure. For example, a given mass of water at its normal boiling point has only $1/1600$ of the volume of the same mass of water vapor under identical conditions. Fig. 10-15 illustrates this disparity in volumes.

figure 10-14  A large quantity of liquid suspended in "free fall" assumes a spherical shape (left). Under the gravitational influence of the earth, liquids assume the shapes of their containers (right).

*Question 10.13*  Using Equation 10-4, and the fact that water has a density of approximately 0.96 g/ml, determine the volumes occupied by one mole of water vapor and one mole of liquid water at 100°C and a pressure of one atmosphere.

The volume occupied by a liquid is affected only slightly by temperature and pressure changes. For example 1.00 liter of liquid water at 20°C and under one atm. pressure will decrease in volume by only 0.0021 liter if the temperature is lowered to 10°C, and by only 0.000049 liter if the pressure is increased to two atm.

*Question 10.14*  Compare these decreases in the volume of liquid water with the decreases in volume of one liter of an ideal gas under the same conditions of temperature and pressure.

Unlike gases, liquids are not always completely miscible with one another in all proportions. This fact is familiar to anyone who has had to vigorously shake a bottle of Italian dressing in order to get the oil and water layers to

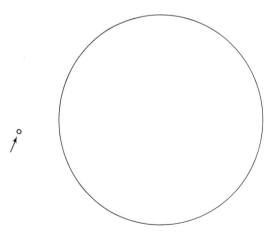

figure 10-15  The tiny water drop on the left will fill a sphere of the diameter shown on the right with water vapor at one atmosphere pressure.

mix together before pouring it on a salad. In general such mixing is only temporary—the liquids soon separate again. In some cases it is possible to mix the liquids so vigorously that the resulting microscopic droplets will not coalesce, and a permanent suspension of one liquid in the other is achieved. This process is known as *homogenization*. As a general rule the greater the similarity in chemical composition the greater the likelihood that liquids will be partially to wholly miscible. For example, gasoline contains several completely miscible liquids that are all hydrocarbons—that is, they contain molecules having only carbon-carbon (C—C) and carbon-hydrogen (C—H) bonds. Water, which contains O—H bonds, is not miscible with gasoline but is completely miscible with small low molecular weight alcohols such as ethanol (grain or drinking alcohol) and methanol (wood alcohol) which also contain O—H bonds.

The phenomena of diffusion and Brownian motion, although less pronounced in liquids, are still observable. A drop of ink in a glass of completely still water will slowly spread out until, after an hour or so, the liquid will be evenly colored. Brownian motion, as mentioned in Section 10.1, was first discovered in liquids. The dance of the tiny particles suspended in a liquid can only be seen with a microscope, since the magnitude of the motions involved is too small to be seen with the naked eye.

Another important property of liquids is *viscosity*. This is the measure of the resistance of a liquid to flow. (Note: Gases also have viscosities, but they are *much* smaller than those of liquids.) Water and gasoline are examples of low viscosity liquids, whereas honey and vaseline have much higher viscosities. Various techniques are used to measure viscosity, one of which involves measuring the rate at which a metal sphere falls in a tube filled with the liquid in question. The lubricating properties of oils and greases are intimately connected with viscosities. A penetrating oil that is used to free a "frozen" bolt must have very low viscosity. On the other hand, engine lubricants must have high enough viscosities to prevent them from being squeezed out from between the moving parts. The viscosities of most liquids decrease markedly with increasing temperature.

*Question 10.15*    Automobile owners used to change to a lower viscosity engine oil in the winter. Now there are so-called "multigrade" lubricants available which make such changes unnecessary, except in very cold climates. What properties should an ideal multigrade lubricant possess?

**The Structure of Liquids.**   Although scientists have not produced a simple ideal model for liquids, they have discovered enough about the microstructure of liquids to provide a basis for at least a qualitative understanding of the observed macroscopic properties.

The average distance between the centers of molecules in liquids is found to be on the order of 2–5 angstroms. (One angstrom, Å, equals $10^{-8}$ cm.) This is about the same spacing that exists in solids. Note that this must be true, since the density of a liquid is about the same as the density of its solid. The interatomic spacing is only slightly greater than the distance at which the electron clouds of the molecules begin to interpenetrate sufficiently to cause repulsive forces to develop. We can therefore think of liquid molecules being almost in "contact." This contrasts with the situation in the gaseous state. At zero degrees Celsius and one atm. pressure, for instance, the average intermolecular distance in an ideal gas is about 33 Å.

*Question 10.16*    Verify this number for the intermolecular distance in ideal gases with the aid of Equation 10-4.

   We see from the two-dimensional representation of a liquid and a gas, as shown in Fig. 10-16, that liquids, unlike gases, do not consist primarily of empty space.
   Despite the cramped quarters, the molecules in the liquid are found to be rather mobile. They move around in a rather chaotic fashion, somewhat like the motion in the gaseous state. However, the average distance traveled between intermolecular collisions (the mean free path) of a molecule in the liquid is only tenths of an Å, as compared to hundreds of Å in gases under ordinary pressures.
   This random motion might lead to the expectation that liquids do not have any regular order or definite structure. Actually, from data obtained using the scattering of x-rays we know that a mapping of the positions of the molecules at any given instant would reveal a high degree of regularity and structure. This order is only "short-range." A two-dimensional representation of what is meant by short-range order is illustrated in Fig. 10-16b. It can be seen that most of the spherical molecules are surrounded by six others providing local order. A few (indicated by shading) are only surrounded by five. These shaded molecules represent imperfections which lead to long-range disorder. It is the continual rapid shifting of the positions of the analogous three-dimensional imperfections within the liquid that permits the molecular

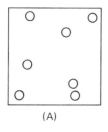

(A)

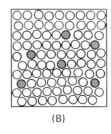

(B)

figure 10-16   Two dimensional models (A) of a gas, and (B) of a liquid.

mobility observed for the liquid state. As the temperature of a liquid is raised, the degree of disorder (and the mobility) increases.

The fact that molecules in a liquid are essentially "touching" one another explains the observed high density, resistance to compression, and low co-efficient of thermal expansion characteristic of the liquid state. The lack of long-range order permits liquids to flow and adopt the shapes of their containers. The chaotic motion of the molecules and short mean free paths are responsible for the microscopically observable Brownian motion of pollen, and other tiny particles suspended in liquids. Scientists are hard at work attempting to extend this qualitative understanding of these and other liquid-phase phenomena to the point where quantitative relationships can be deduced.

## 10.3  crystalline solids

The solid state is characterized by an obvious structural rigidity that contrasts with the fluidity of gases and liquids. Solids have well defined volumes and shapes and do not conform to the shapes of containers in which they are placed. The effect of pressure and temperature changes on the volume of a solid is generally even smaller than is the case with liquids.

We shall confine our discussion to crystalline or "true" solids. Other materials such as glass, certain plastics, and asphalt exhibit some of the properties of the solid state but they are more precisely classified as falling into the fuzzy boundary between solids and liquids alluded to in Chapter 1.

Studies of the macroscopic properties of solids date back to earliest recorded history. The beauty and symmetry of crystalline gemstones attracted the interest of the ancient philosophers who recorded the various forms in which minerals were found. Various theories and superstitions were advanced, such as the ancient Greek, and Roman, belief that quartz resulted from the exposure of ice to intense cold. It was not until 1669, however, that the true foundation was laid for the scientific study of crystals. In that year a Danish professor of anatomy, Nick Stensen (also known as Nicholas Stena in his role as court physician to the Duke of Tuscany) published a study in which he advanced the proposition, based on careful measurement, that although different crystals of quartz may appear to have different external shapes and sizes, the *angles* between *similar faces* are *always the same*. With the invention of more precise tools for the measurement of crystal angles in the late 18th century came the extension of the principle of the constancy of interfacial angles—sometimes termed the "first law of crystallography"—to other crystalline substances. The reason for the different external appearance is due to the fact that during their formation the rates of growth of corresponding faces of crystals grown under different conditions may differ. (See Fig. 10-17.)

A theoretical explanation for the regularities exhibited by the shapes of

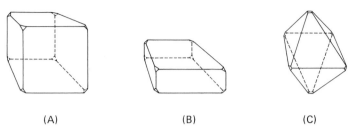

(A)                              (B)                              (C)

figure 10-17   Shapes of sodium chloride crystals vary according to
the conditions under which they are grown. (After *University
Chemistry*, Second Edition, by B. H. Mahan. 1969. Addison-
Wesley.)

crystals was first proposed in 1784 by Abbé René Just Haüy, a professor at
the University of Paris. He suggested that the external symmetry was a
consequence of the regularity of the internal arrangement of the building
blocks of the crystal which he believed to be tiny polyhedra. The development
of x-ray scattering analysis during the first quarter of this century has resulted
in the confirmation of the essentials of Haüy's theory. Crystalline regularity
is indeed a reflection of an ordered microstructure, consisting of molecules
or atoms arranged in definite geometric patterns. Molecules in a solid are still
in constant motion, but this motion is restricted to small oscillations about
fixed equilibrium positions (see Fig. 10-18).

**The Crystal Lattice.**   The repeating three dimensional array of atoms,
molecules, or ions that describes the internal structure of a true solid is called
a *crystal lattice*. The basic building block of the structure is called the *unit*

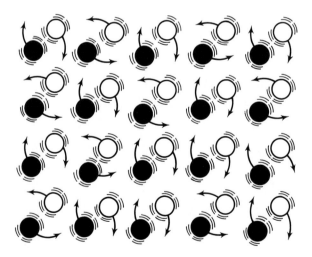

figure 10-18   Restricted motion of molecules in a crystal.

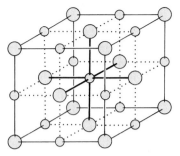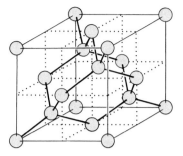

figure 10-19   Unit cells of common salt (NaCl) (left), and of the diamond form of carbon (right).

*cell,* and the entire lattice can be built up by a three-dimensional repetition of these structural units. Fig. 10-19 illustrates examples of such unit cells for common salt (NaCl) and for the diamond form of carbon.

*Question 10.17*   Although the picture of the unit cell of NaCl shows the position of 27 ions (14 $Na^+$ and 13 $Cl^-$) the cell is said to "contain" only 8 ions (4 $Na^+$ and 4 $Cl^-$). Can you explain this apparent contradiction? (Hint: most of the ions in the cell are shared by adjacent cells in the lattice.)

Crystal lattices can be classified according to the relative lengths of the edges of a unit cell and their angles of intersection at a corner of a unit cell. It turns out that there are only seven different unit cell geometries that can be accommodated in a space-filling three-dimensional repeating structure, and these result in seven crystal systems. If we designate the three edges as $a$, $b$, and $c$, and the angles as $\alpha$, $\beta$, and $\gamma$ as illustrated for the cubic system in Fig. 10-20, we can tabulate the characteristics of the seven systems as shown in Table 10-2 and depicted in Fig. 10-21.

**Types of Solids.**   Of more practical significance than the crystal system

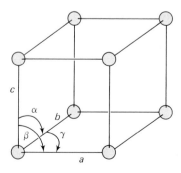

figure 10-20   Designation of three edges $(a,b,c)$, and the angles $(\alpha,\beta,\gamma)$ of intersection, of the unit cell (illustration for the cubic crystal system).

table 10.2    the seven crystal systems.

| System | Axial Lengths | Crystal Angles | Examples |
|---|---|---|---|
| Cubic | $a = b = c$ | $\alpha = \beta = \gamma = 90°$ | NaCl; CsCl; Cu |
| Tetragonal | $a = b; c$ | $\alpha = \beta = \gamma = 90°$ | $SnO_2$; Sn(white) |
| Orthorhombic | $a; b; c$ | $\alpha = \beta = \gamma = 90°$ | $K_2SO_4$; $I_2$ |
| Rhombohedral° | $a = b = c$ | $\alpha = \beta = \gamma \neq 90°$ | $Al_2O_3$; As; Bi |
| Monoclinic | $a; b; c$ | $\alpha \neq \beta = \gamma = 90°$ | $KClO_3$; S |
| Triclinic | $a; b; c$ | $\alpha \neq \beta \neq \gamma \neq 90°$ | $K_2Cr_2O_7$ |
| Hexagonal | $a = b; c$ | $\alpha = \beta = 90°; \gamma = 120°$ | C(graphite) |

° The rhombohedral class is sometimes considered to be a special case of the hexagonal system rather than a separate crystal system by some crystallographers. Other crystallographers use the name trigonal rather than rhombohedral to designate this class.

classification of a solid is the characterization of crystals in terms of the types of forces that bind them together. Four general types of crystals distinguishable on this basis are molecular, covalent, ionic, and metallic.

*Molecular solids* consist of individual molecules or atoms held together by van der Waals' forces and, in the case of polar molecules, by dipole-dipole interactions. Examples are iodine, sulfur, dry ice (solid $CO_2$), and solid inert

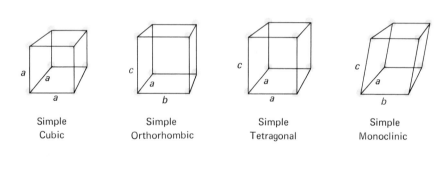

Simple Cubic    Simple Orthorhombic    Simple Tetragonal    Simple Monoclinic

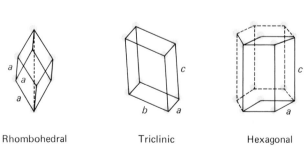

Rhombohedral    Triclinic    Hexagonal

figure 10-21    Unit cells of the seven crystal systems. (From *University Chemistry* Second Edition, by B. H. Mahan. 1969. Addison-Wesley.)

gases. The relatively weak intermolecular forces explain the low melting points observed for crystals of this type. Also, since the force is usually not highly directional, the molecules are not very resistant to small shifts in position and as a result molecular crystals are usually soft. An important subset of molecular crystals contains those in which a more specific type of interaction called *hydrogen bonding* plays a significant role in holding the crystals together. Hydrogen bonding is a phenomenon that involves the interaction of a highly polar bond involving a hydrogen atom (such as O—H or F—H) with an atom that bears a partial negative charge (such as another O or F atom). Water is the most common substance for which hydrogen bonding is a very important factor. (See Fig. 10-22.) The structure of ice (Fig. 10-23) involves extensive hydrogen bonding. Since hydrogen bonds are considerably stronger than van der Waals' interactions, hydrogen bonded solids tend to melt at much higher temperatures than other comparable molecular solids. The hydrogen bonds also have preferred directions, resulting in crystals that are harder, and which tend, when struck, to shatter rather than to distort.

*Question 10.18*   Look up the melting point of $H_2S$ and compare it to that of $H_2O$. Explain why this comparison suggests that hydrogen bonding is not an important factor in the structure of solid $H_2S$.

Covalent bonding results in the second major structural class of crystals. In these *covalent solids* strong bonds link all the atoms in the crystal together into what might be considered one giant molecule. Diamond (see Fig. 10-19) and quartz ($SiO_2$) are two important examples of covalent solids. Since the bonds are very strong and extend in fixed directions relative to one another, crystals of this type are the hardest materials known and have very high melting points.

A third type of solid, *the ionic*, consists of positive and negative ions held together by strong electrostatic forces. Common examples are table salt (NaCl), saltpeter ($KNO_3$), alum ($KAl(SO_4)_2$), and baking soda ($NaHCO_3$). Ionic solids tend to have high melting points. The structures of most ionic solids are highly resistant to arbitrary distortions that tend to produce electric repulsions between similarly charged ions. Thus, these crystals are usually brittle and

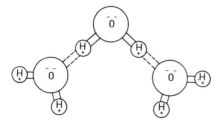

figure 10-22   Hydrogen bonding in water. In addition to the strong polar O—H bonds, weaker O···H hydrogen bonds are formed between the negative oxygen atom and the positive hydrogen atoms.

shatter by cleaving along well defined planes in the crystal, which contain roughly equal number of positive and negative charges. Section 2 of the Handling the Phenomena section for this chapter describes how to locate the cleavage planes in a simple crystal.

*Question 10.19*     Graphite (see Fig. 10-24) consists of sheets of carbon atoms. The bonding *within* a sheet is covalent. The bonding between the sheets is due to van der Waals' forces. Predict the properties of graphite and explain why it can be used as a effective lubricant in machinery that operates at high temperatures.

The final category, *metallic solids,* can be thought of as a lattice of positively charged metal ions held together by a gas of electrons. The valence electrons

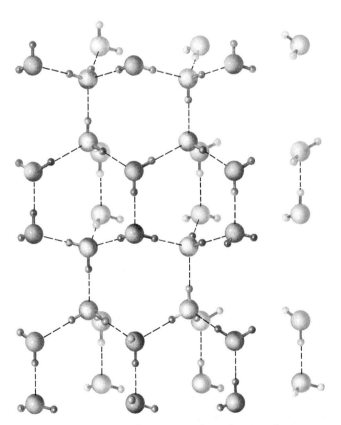

figure 10-23    The crystal structure of ice showing the importance of hydrogen bonds (dashed lines). (From *General Chemistry,* Third Edition, by L. Pauling. Copyright © 1970. W. H. Freeman and Co.)

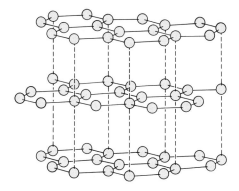

figure 10-24    The structure of graphite. Covalently bound, two-dimensional, hexagonal sheets of carbon atoms are held together in the third dimension by weak molecular bonds. (From *Chemistry: An Experimental Science*, G. C. Pimentel, ed. Copyright © 1963. W. H. Freeman and Co.)

are delocalized over the entire crystal, rather than belonging to individual metal atoms. This mobility of electrons results in the high electrical and thermal conductivity of metals. Predicting the other properties of metallic crystals is complicated by the fact that covalent bonding is also present to varying degrees in some metallic crystals. Thus the melting points of metals vary from very low (mercury melts at $-38.9°C$) to very high (tungsten melts at $3380°C$) and the crystals can be soft and malleable, or hard and brittle.

A summary of the characteristics of the four types of crystals is given in Table 10-3.

**Defects in Crystals.**   Actual crystals deviate somewhat from the idealized perfect lattice. Various types of structural defects occur. These include holes in the lattice, known as lattice vacancies, the presence of extra interstitial impurity atoms or ions, and lattice dislocations. These are illustrated in Fig. 10-25. Structural defects tend to increase as the crystal approaches its melting

table 10.3    types of crystalline solids.

| Type | Lattice Sites Occupied by | Forces Holding Crystal Together | Properties |
|---|---|---|---|
| Molecular | Molecules | van der Waals, dipole-dipole, hydrogen bonding | Low-melting point and soft |
| Covalent | Atoms or groups of atoms | Covalent bonding | Very high-melting point and very hard |
| Ionic | Ions | Electrostatic interaction | High-melting point and brittle |
| Metallic | Postive ions | Attraction between positive ions and "electron gas" and sometimes covalent bonding | Variable melting point and hardness. High electrical and thermal conductivity |

Vacancy          Interstitial atom          Dislocation

figure 10-25   Three types of crystal defect.

point. Crystals that are formed by the rapid cooling of a liquid generally have many structural defects frozen in.

Another class of defects is due to the presence of impurities. Very small levels of impurity can drastically alter such properties as the electrical conductivity and strength of solids. The controlled introduction of impurities to produce materials with specific desirable properties is exemplified by the manufacture of semiconductors for use in solid state electronic devices.

## 10.4   changes of state

Water is the only substance that people frequently encounter in all three physical states. We will begin our consideration of the changes of state by describing the macroscopically observable changes that occur for this common substance, and then proceed to consider how the general phenomenon of transitions from one state to another can be understood in terms of the microstructure of matter.

Suppose we place a piece of ice in a beaker and put it under a bell jar (Fig. 10-26). Next we place the apparatus in a freezer at $-10°C$. As long as we maintain the system at this temperature no apparent change occurs.

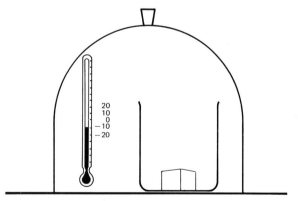

figure 10-26   Ice in an enclosure at a temperature of $-10°C$ will produce a water vapor pressure of $2.8 \times 10^{-3}$ atmosphere.

All the water *seems* to remain in the solid form. As a check on this assumption let us remove a sample of the air in the jar and test it for water content. We find that there *is* a small partial pressure of water in the air equal to $2.8 \times 10^{-3}$ atm. Adding more ice to the beaker does not change the water content of the air because the air is *saturated* with water. This partial pressure, which is called the *vapor pressure* of the ice, is observed to be fixed by the temperature. If we now remove the bell jar from the freezer and allow the system to warm slowly and monitor the water vapor pressure we find that it rises slowly until it reaches $6.0 \times 10^{-3}$ atm. at $0°C$. At this point the ice in the beaker is observed to begin to melt. The temperature of the ice-water mixture and the vapor pressure during the melting process remain fixed at $0°C$ and $6.0 \times 10^{-3}$ atm. respectively until *all the ice has melted.* At this point the temperature and vapor pressure of the water again begin to rise. Eventually the system reaches room temperature, which happens to be $23°C$, and the vapor pressure is found to level off at $2.8 \times 10^{-2}$ atm. In order to continue to warm the system we transfer it to an oven and slowly heat the system. The temperature and vapor pressure continue to rise until at $100°C$ the vapor pressure equals $1.0$ atm. At this point the water begins to boil. If we do not remove the bell jar at this point, the vapor pressure and temperature will continue to rise as the liquid boils. If, however, the jar is removed, allowing the liquid to remain in contact with the atmosphere, we find that the liquid remains at $100°C$ until all of the water has vaporized.

Consider another experiment. This time we take a sample of laboratory air and carefully measure both its temperature, and the partial pressure of water in it, which we find to be $23.0°C$ and $1.40 \times 10^{-2}$ atm. respectively.

*Question 10.20*   Why is the partial pressure of water in the laboratory air not necessarily equal to the vapor pressure of water at $23°C$, which was found to be $2.8 \times 10^{-2}$ atm.?

Next, we take another sample of laboratory air and seal it in a glass bulb. We then dip the bulb into a water bath and slowly cool the bath by adding small amounts of cold water. We remove the bulb after each small lowering of its temperature. At $11°C$ we notice that a slight film of water has condensed on the inside of the sealed bulb. We look up the vapor pressure of water at $11°C$ and find that it is $1.29 \times 10^{-2}$ atm. which is slightly less than the partial pressure of water that would have remained in the sample of air if no condensation had occurred. We have cooled the air in the bulb below the point where it was completely saturated with water, and therefore the water has condensed. The ratio of the partial pressure of water in air to its saturated vapor pressure at that temperature is called the *relative humidity* of the air.

*Question 10.21*   What was the humidity in the laboratory when this experiment was performed?

*Question 10.22*    Why are human beings more uncomfortable on a day when the temperature is 20°C if the humidity is 98%, rather than if it is 40%?

**The Kinetic Theory of Change of State.**    In Chapter 7 we established the proportionality between absolute temperature and the average kinetic energy of molecules. In Section 1 of this chapter we demonstrated how the kinetic theory can be used to derive the observed properties of gases. The temperature of solids and liquids is also a measure of the kinetic energy of the molecules. In fact, when we say that any two objects are at the same temperature, what we mean, in terms of the microstructure of matter, is that they have not only the same *average* molecular kinetic energy, but that this energy is *distributed* in the same way. All of the molecules do not have the same energy at any given temperature, and what we mean by the distribution of energy is illustrated by the curves in Fig. 10-27 where the fraction of molecules is plotted against the kinetic energy. The solid curve (T = 500°K) represents a higher temperature energy distribution than the dashed curve (T = 300°K). We see that the general shape of the curve becomes more flattened and shifted as the average kinetic energy shifts to higher values with increasing temperature.

At −10°C ice molecules will be oscillating about their equilibrium positions with average kinetic energies of vibration represented by the appropriate curve for that temperature. As we can see from Fig. 10-27 some small fraction of the molecules will have kinetic energies much higher than the average. If these high energy molecules are at the surface of the solid they can escape

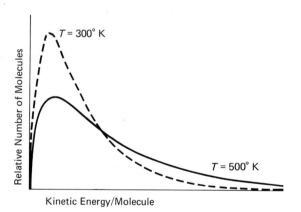

figure 10-27    The distribution of kinetic energy among molecules for two different temperatures. (From *Principles of Physical Science*, Second Edition, by F. T. Bonner, M. Phillips and J. Raymond. 1971. Addison-Wesley.)

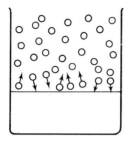

figure 10-28   When the equilibrium vapor pressure has been reached the rate of condensation equals the rate of evaporation.

from the potential energy well and enter the gaseous state. This mechanism is responsible for the presence of water in the air above the ice. After a short time, due to collisions with air molecules and with the walls of the container, these escaped water molecules will again be in thermal equilibrium with the surroundings and will have an energy distribution that conforms to the curve for $-10°C$. As a water molecule bounces around inside the container it will eventually strike the surface of the ice. When this happens it will be subjected to being trapped in the potential wells from which other molecules have broken loose. The likelihood is that it will not have a high enough energy to escape, and it will therefore become forzen into the solid. This constant evaporation and condensation of molecules soon leads to a *dynamic equilibrium* in our closed container. This happens when the rate at which molecules are escaping from the surface is equal to the rate at which they are becoming trapped again. The escaping and trapping rates are fixed by the fraction of molecules that have energies higher or lower than the minimum escape energy at any given temperature. When a sufficient concentration (pressure) of gaseous molecules has built up to make the condensation rate equal to the evaporation rate, the *equilibrium vapor pressure* has been established as shown in Fig. 10-28.

As the temperature of the ice rises, the vapor pressure rises because a larger fraction of the molecules will have high energies. A very high energy molecule inside the crystal will occasionally break out of its equilibrium potential well (Fig. 6-17). This will lead to a defect in the crystal structure. Increasing temperature will then lead to an increase in the percentage of defects. At $0°C$ we have reached the point at which the percentage of defects, although still quite low, is too high to permit the long range order required for the stability of the crystal. This causes the collapse of the lattice structure of the solid, and melting occurs. In the liquid state the molecules are still subject to forces that are strong enough to keep them essentially in contact, but they have achieved mobility by virtue of the breakdown of the highly rigid ordered crystal structure. This change of environment will result in a change in the average potential energy of the molecule. The solid-liquid transition always involves an *increase* in the potential energy. Heat flowing into the ice-water system at $0°C$ (or into any other solid-liquid system at its melting point) goes to supply the potential energy needed in order to accomplish this change of state. This potential energy of melting is called the *heat of fusion*. For ice it is 80 calories per gram. Since the solid and liquid have the same distribution of kinetic energies no change in temperature occurs during the melting process.

Once melting has occurred, heat energy can again cause a continual increase in temperature and vapor pressure as the kinetic energy distribution shifts toward higher energies. When we reach 100°C the vapor pressure has climbed to one atmosphere. If the liquid is in contact with the atmosphere the vapor pressure is now the same as the external pressure on the liquid surface. This means that the molecules escaping from the liquid can exert sufficient pressure to develop bubbles inside the liquid. It is this phenomenon that we call boiling. If a pressure of one atmosphere is maintained on the water surface all of the energy that we feed in at 100°C will go to convert liquid water to gaseous water, which involves an even larger potential energy increase than the solid to liquid transition. Since no change in the kinetic energy distribution occurs during boiling under constant pressure the temperature will remain constant. The energy associated with the liquid-gas transition is called the *heat of vaporization*. For water it is found to be 540 calories per gram.

*Question 10.23*    Why does the temperature of water in a *closed* container continue to increase if heat is supplied while it is boiling?

*Question 10.24*    Look up the boiling point of hydrogen sulfide, $H_2S$. Do you think that hydrogen bonding occurs to any great extent in liquid water? Explain.

*Question 10.25*    Why is the heat of vaporization of a substance greater than its heat of fusion?

**Phase Diagrams.**    If a pure substance is put in a cylinder that has one end sealed with a movable piston, as shown in Fig. 10-29, we can study its behavior under various conditions of temperature and externally applied pressure. The results of such studies can be illustrated by what is known as a phase diagram, which is a graphical representation of the state of the substance as a function of temperature and pressure.

Phase diagrams for water and carbon dioxide are given in Fig. 10-30. The solid, liquid, and gaseous regions are separated by lines that represent an equilibrium between two of the phases. All three phases are found to coexist at only one point, which is called the *triple point*. Such diagrams can be used to determine what changes will occur at constant pressure if the temperature is changed. For example, if the dotted line representing one atmosphere on

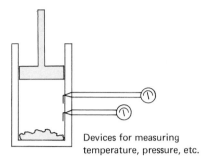

Devices for measuring temperature, pressure, etc.

figure 10-29   A pure substance in a cylin- der enclosed by a piston.

the water diagram is traced from left to right it is seen to cross the solid-liquid line at 0°C and the liquid-gas line at 100°C, which are the normal melting and boiling points, respectively. A similar procedure for $CO_2$ reveals that, in this case, only the solid-gas line is crossed. This means that $CO_2$ will not melt at one atmosphere, as anyone who has seen dry ice *sublime* without melting already knows. We see further that if dry ice is heated while being kept at a pressure of more than 5.1 atmospheres it *will* melt before it vaporizes.

*Question 10.26*   Below what pressure would ice be observed to transform directly from the solid state into the gaseous state without melting?

Another fact revealed by the phase diagrams is that water melts at lower temperatures when the pressure is increased. It is because of this phenomenon that ice skating is made possible. The pressure of the skate blade melts a thin

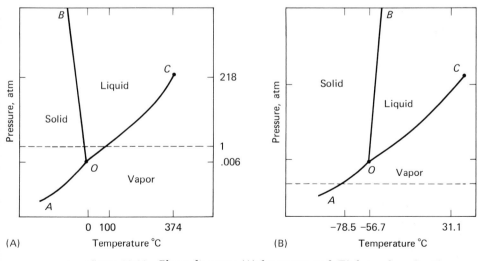

figure 10-30   Phase diagrams (A) for water, and (B) for carbon dioxide.

film of water underneath, allowing almost frictionless gliding. The opposite behavior is observed for $CO_2$ and most other substances. This peculiar property of water is a direct consequence of the fact that ice is *less* dense than water at the freezing point, which, in turn, is due to the unusually open hydrogen bonded structure of water. The fact that ice is less dense than water has important consequences for the fish that swim in our lakes and rivers. When these bodies of water freeze, the ice floating on top acts as an insulating layer (ice is a poorer conductor of heat than water) and prevents the water below from freezing. If the ice did not float, many of our lakes and streams would freeze solid, trapping any life that they contained.

Some substances can exist in more than one solid form, depending on the temperature and pressure. This is illustrated by the phase diagram (Fig. 10-31) for sulfur in which the rhombic and monoclinic solid forms appear. Two different solid forms of the same element are called *allotropes*. Solid-solid transitions frequently occur very slowly, and some allotropes that are actually unstable at room temperature can thus appear to be completely stable. If this were not the case diamond jewelry would not be very popular since graphite is the stable form of crystalline carbon at room temperature and atmospheric pressure! A solid-solid transition that *has* caused trouble is one involving the two allotropes of tin. Metallic or "white tin" is stable at temperatures above 13.2°C, whereas a non-metallic form, known as gray tin, is stable at lower temperatures. Organ pipes in old European cathedrals used to be made of tin. These cathedrals lacked central heating, and those in the north that were exposed to temperatures below 13°C for long periods of time developed "tin

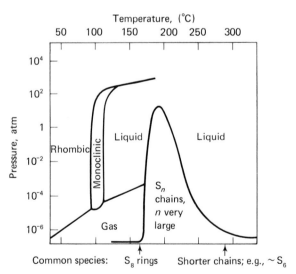

figure 10-31    Phase diagram for sulfur, a substance that has more than one solid form.

disease" and slowly crumbled. This phenomenon, which previously had been attributed to various supernatural causes, is now known to be the result of the slow change from metallic to non-metallic tin.

## summary

In this chapter, the general characteristics of the gaseous, liquid, and solid states were investigated.

We found that the empirical volume-pressure-temperature relationship for any gas is given, to a first approximation, by the gas law, $PV/T = c$. Furthermore, the observation that equal volumes of any two gases at the same temperature and pressure contain equal numbers of molecules permits the replacement of the constant $c$ by the product of the number of moles of gas, $n$, and a universal constant, $R$. The use of this equation in various quantitative problems was demonstrated, including the use of the method suggested by Cannizzaro for the determination of atomic weights. We completed our empirical investigation of gaseous properties by noting the additive nature of gas pressures (Dalton's Law of partial pressures), the inverse proportionality between the diffusion rate of a gas and the square root of its molecular weight (Graham's Law) and the erratic motion of smoke particles suspended in a gas (Brownian motion).

We next turned to the development of the kinetic model of gases which underlies our present understanding of gas-phase phenomena. This model pictures gases as consisting of widely separated molecules that are constantly moving in random straight line paths. Intermolecular forces are negligible except during perfectly elastic collisions. Starting with these assumptions and the postulated proportionality between average molecular kinetic energy and absolute temperature (see Chapter 7), and making a few reasonable approximations, we were able to derive the universal gas law for the particular (but generalizable) case of a gas confined to a cube.

We concluded our study of gases by examining the sources of deviations from ideal behavior that occur at high pressures, and low temperatures.

Turning our attention to liquids we observed several characteristic properties. Liquids do not expand to fill any available volume. They are much more dense than gases. Liquid volume changes only slightly with temperature or pressure. Liquids are not all miscible with one another in all proportions. Liquid viscosities (resistance to flow) can differ greatly from one another and generally decrease rapidly with temperature.

Unfortunately, no universal relationship such as the ideal gas law has been discovered for liquids. A model for the liquid state resembles the gaseous model only in that molecules are in constant motion. Liquid molecules are in very close contact but are still rather mobile. Despite this mobility the positions of molecules in a liquid reveal a high degree of short-range order.

The local regularity characteristic of the structure of liquids develops into long-range, rigid, highly-ordered structures when the liquid becomes a crystalline solid. The repeating geometrical patterns of molecules which constitute the microstructure of crystalline matter give rise to the regularities and symmetries observed in the external shapes of crystals. Molecules in crystals are much more confined than in gases or liquids, but are still free to oscillate about their fixed equilibrium positions.

We examined the classification scheme which divides crystals into seven systems based on the shape of the basic repeating structural building block called the unit cell. Another important classification scheme for crystals was described in terms of the forces holding the crystal together. Molecular, covalent, ionic, and metallic crystals were contrasted. Variations in such properties as melting point, hardness, and electrical conductivity were attributed to the differences between the types of intermolecular forces involved.

Having investigated the properties of the three common states of matter, we then considered the transitions from one state to another. Again we began by empirically examining physical properties (temperature and vapor pressure). In this case it is the changes in these properties that accompany changes in phase that were looked at. We found that both the temperature and vapor pressure remain constant during phase changes. To explain these observations the kinetic model of a gas was expanded to include liquids and solids, and the notion of a distribution of molecular kinetic energies was introduced. In terms of this model vapor pressure was explained as resulting from a temperature-dependent dynamic equilibrium between evaporating and condensing molecules. Phase changes were characterized as involving changes in potential energy, rather than by changes in kinetic energy. Since temperature is postulated to be a measure of the average molecular kinetic energy we have thus explained why the temperature stays fixed during phase changes.

Finally, phase diagrams were introduced and examined as a graphical representation of the state of a substance under varying conditions of temperature and pressure. We used these diagrams to understand why dry ice evaporates without melting under atmospheric conditions and why ice floats on top of water.

*Answers to Questions*    10.1    To convert Equation 10-2 to the Celsius scale first recall the relationship:

$$T(^\circ K) = t(^\circ C) + 273$$

Substituting in Equation 10-2 we see that

$$V = b(t(^\circ C) + 273)$$

or    $V = bt + 273b$

This is an equation for a straight line with slope $b$ and intercept $273b$.

10.2   We can rewrite Equation 10-3 in the form:

$$P = \frac{Tc}{V}$$

At constant temperature we can replace the product $Tc$ by the single constant, $a$, giving $P = a/V$ which is Equation 10-1. Similarly, if we rewrite Equation 10-3 in the form:

$$V = \frac{Tc}{P}$$

and consider the relationship between $V$ and $T$ at constant $P$ we can replace the constant $c/P$ by the single symbol, $b$, giving $V = bT$, which is Equation 10-2.

10.3   By rearranging Equation 10-4 we can obtain:

$$V = \frac{nRT}{P}$$

The molecular weight of oxygen gas ($O_2$) is 32.0. Therefore 32.0 grams equals 1.00 mole. Substituting the numerical values for $n$, $R$, $T$, and $P$ gives:

$$V = \frac{(1.00)(.0821)(273)}{1.00} = 22.4 \text{ liters}$$

10.4   The product gas contains the same number of molecules as there are in the starting sample of $N_2$ since their volumes are equal. Therefore each product molecule must contain two atoms of nitrogen. The volume of oxygen used is twice that of the product. Therefore, two $O_2$ molecules must combine with each $N_2$ molecule to form the product. The resulting molecule contains four atoms of oxygen and two of nitrogen, and must have the formula $N_2O_4$.

10.5   Using Equation 10-7 we find

$$\delta_{(A)} = \frac{(1.00)(71.0)}{(.0821)(373)} = 2.32 \text{ g}/\ell$$

$$\delta_{(B)} = \frac{(1.00)(119)}{(.0821)(373)} = 3.90 \text{ g}/\ell$$

$$\delta_{(C)} = \frac{(1.00)(154)}{(.0821)(373)} = 5.04 \text{ g}/\ell$$

$$\delta_{(D)} = \frac{(1.00)(85.0)}{(.0821)(373)} = 2.78 \text{ g}/\ell$$

$$\delta_{(E)} = \frac{(1.00)(117)}{(.0821)(373)} = 3.84 \text{ g}/\ell$$

10.6   If compounds always contained elements in multiples of two atoms (2, 4, 6, etc.) then Cannizzaro's method would give twice the actual atomic weight. If, however, compounds always contained at least two atoms, but could contain odd numbers as well, then it would be evident that the common divisor would still be the actual atomic weight!

10.7   The lecturer might open a bottle of a smelly compound of known molecular weight and record the time it takes for the molecules to diffuse to the students in the front row of seats. A second odoriferous substance of much higher molecular weight could then be used and a much slower diffusion rate (longer time for front row students to detect the smell) would be recorded.

10.8   The inelastic nature of collisions between objects is due to the conversion of kinetic energy to internal energy. An increase in *internal* energy means that the molecules *within* the object have had their kinetic and/or potential energies raised. It is reasonable to postulate that individual gas molecules cannot continuously absorb internal energy.

10.9   The kinetic theory assumes that there is no interaction between gas molecules except for collisions. Thus, adding two gases together in a container will result in an average rate of collisions with the walls equal to the sum of the rates for the two separate gases in the same size container. The momentum change per unit time, and therefore the pressures, will thus be additive. This is what Dalton discovered.
   Brownian motion is due to the constant bombardment of extremely small particles by the rapidly moving gas molecules. For a very small particle, the number of molecules striking one side in a given instant will not be exactly equal to the collisions on its opposite side. This small imbalance can be used to predict the erratic motion that is observed.

10.10   Only water of $110°C$ will show large negative deviations. Oxygen at $50°C$ and helium at $-50°C$ are both far above the point where they can be liquified.

10.11   *At* $50°C$

Calculation using Equation 10-27:

$$\left(P + \frac{(1.00)^2(1.390)}{(1.00)^2}\right)(V - (1.00)(.03913)) = (1.00)(.0821)(323)$$

$$(P + 1.390)(1.00 - .03913) = 26.5$$

$$(P + 1.390)(.961) = 26.5$$

$$.961P + (.961)(1.39) = 26.5$$

$$P = \frac{26.5 - 1.34}{.961}$$

$$PV = 26.2 \; \ell \cdot atm$$

Calculation using Equation 10-4:

$$\frac{PV}{T} = nR$$

$$\frac{PV}{323} = (1.00)(.0821)$$

$$PV = 26.5 \; l \cdot atm$$

*At* $-100\,^{\circ}C$

Equation 10-27:

$$.961P + 1.34 = (1.00)(.0821)(173)$$

$$P = \frac{14.2 - 1.34}{.961}$$

$$PV = 13.4 \; l \cdot atm$$

Equation 10-4:

$$\frac{PV}{173} = (1.00)(.0821)$$

$$PV = 14.2 \; l \cdot atm$$

10.12   Bromine and mercury are the only elements that are liquids at *room temperature* and one atmosphere. Perhaps you have encountered pure glycerin; ethyl, methyl, and isopropyl alcohol; mercury; or a few other pure compounds that are liquids at room temperature. (*Note:* gasoline, turpentine, "rubbing" alcohol and most other fluids are solutions or mixtures of liquids.)

10.13   For one mole of gaseous water at $100\,^{\circ}C$ and 1 atm:

$$V = \frac{nRT}{P} = \frac{(1.00)(.0821)(373)}{1.00} = 30.6 \; l$$

For one mole of liquid water at $100\,^{\circ}C$ and 1 atm:

$$V = \frac{18 \text{ g}}{0.96 \text{ g}/ml} = 19 \text{ m}l = .019 \; l$$

10.14   For one liter of an ideal gas at $20\,^{\circ}C$ and $1.00$ atm a decrease in temperature to $10\,^{\circ}C$ will cause a decrease in volume to

$$V = \left(\frac{283}{293}\right)(1.00) = 0.966 \; l$$

and an increase in pressure to $2.00$ atm will cause a decrease in volume to

$$V = \left(\frac{1}{2}\right)(1.00) = .50 \; l$$

**10.15** A multigrade lubricant should have a low enough viscosity when the engine is cold to allow the engine to turn over freely. But, it should also have a high enough viscosity when the engine is hot to maintain a lubricating coating on the rapidly moving parts. It should also be resistant to chemical breakdown at the high operating temperatures of the engine.

**10.16** At 0°C and 1 atm pressure 1 mole of gas will occupy:

$$V = \frac{nRT}{P} = \frac{(1.00)(.0821)(273)}{(1.00)} = 22.4 \text{ liters}$$

One liter is $1000 \text{ cm}^3$. Therefore $6.02 \times 10^{23}$ molecules will be distributed in $22,400 \text{ cm}^3$ of space. Consider a cube of this volume. One edge will be $\sqrt[3]{22,400} = 28.1$ cm long. There will be $\sqrt[3]{6.02 \times 10^{23}} = 8.45 \times 10^7$ molecules spaced out along this edge. The average distance between molecules will therefore be

$$\frac{28.1 \text{ cm}}{8.45 \times 10^7} = 3.3 \times 10^{-7} \text{ cm}$$

$$= 33 \times 10^{-8} \text{ cm}$$

$$= 33 \text{ Å}$$

**10.17** The eight $Na^+$ ions at the *corners* of the unit cell in Fig. 10-17 are each shared by eight cells. Thus only $\frac{1}{8}$ of eight, or one, of these atoms is "in" the unit cell. Similarly each $Na^+$ ion in the center of the six cube faces is shared by two cells, giving $\frac{1}{2}$ of six, or three, more $Na^+$ ions to the cell.

For the $Cl^-$ ions, we find twelve that are located on the edges of a cell, each of which is shared by four cells, providing $\frac{1}{4}$ of twelve, or three, for the cell in question. The remaining $Cl^-$ ion is in the center of the unit cell and is not shared with any other. Thus we have four $Na^+$ and four $Cl^-$ ions per unit cell.

**10.18** The melting point of $H_2S$ is $-83°C$ whereas that of water is $0°C$. The relatively low melting point of $H_2S$ indicates that its crystal structure is much more easily destroyed than that of ice. This suggests that an additional attractive force between molecules, such as that due to hydrogen bonds, exists in solid $H_2O$, but not in solid $H_2S$.

**10.19** Graphite consists of sheet-like layers which have considerable two-dimensional strength due to covalent bonds, but which can readily slip and slide over one another due to the weak bonds between adjacent sheets. This slipping of the molecular sheets at high temperatures and pressure is responsible for the lubricating properties of this form of carbon.

**10.20** The air in the laboratory does not necessarily have to be *saturated* with water vapor.

**10.21** $\text{Humidity} = \dfrac{1.4 \times 10^{-2}}{2.8 \times 10^{-2}} = 0.50 \text{ or } 50\%$

**10.22** At 98% humidity the air is nearly saturated with water and the rate of evaporation of perspiration will be slow. Humans depend on rapid evaporation of perspiration to keep their bodies feeling comfortable.

10.23   The pressure in the system will increase, thus increasing the boiling point.

10.24   The normal boiling point of $H_2S$ is only $-62°C$, whereas that of liquid water is $100°C$. This difference is ascribable to a high degree of hydrogen bonding in liquid water, which results in a much greater energy requirement for vaporization than is found for $H_2S$, which is assumed to involve little or no hydrogen bonding.

10.25   Vaporization involves taking molecules that are almost touching and increasing their distances to the point where attractive internuclear forces are almost nil (as illustrated by Fig. 10-13). Fusion involves only a small change in average internuclear distance and a relatively small change in potential energy.

10.26   Below its triple point pressure, which is 4.58 torr.

### handling the phenomena

### 1.   *Molecular Weights of Gases*
**CAUTION!** If an open flame from a bunsen burner rather than a hot plate is used as a source of heat, make sure that a fire extinguisher is close at hand.

Perform this experiment in a well-ventilated area. Excessive inhalation of ethylene dichloride vapor can be harmful and should be avoided.

Section 10.1 illustrated how the gas law can be used to relate molecular weights to gas densities (see Equation 10-8). This relationship is the basis for several sophisticated techniques for making accurate measurements of the molecular weights of volatile compounds. The following, relatively crude, adaptation of this technique is suitable for measuring the molecular weights of gases that are at least twice as dense as air to about 10% accuracy.

Using a triple beam balance, measure the total mass of a clean, dry 250 ml Erlenmeyer flask, a $5 \times 5$ cm square of aluminum foil, and a rubber band. Add 2 or 3 ml of ethylene dichloride (also called 1,2-dichloroethane) to the flask. Cover the flask with the aluminum foil as tightly as possible and secure the foil with the rubber band. Make a pin hole in the foil cover. Now wrap a second, larger piece of aluminum foil over the first piece. Place the flask in a 1 liter beaker which has about 50 ml of water in it. Cover the beaker with a watch glass and bring the water to a boil. The ethylene dichloride in the flask will slowly vaporize and its dense vapor will drive out the less dense air. The two or three ml of liquid is more than enough to fill the flask with vapor. The excess will escape through the hole in the foil. When the last trace of liquid disappears, the flask will be full of ethylene dichloride vapor at $100°C$ ($373°K$). Carefully remove the flask from the beaker and allow it to cool to room temperature. The ethylene dichloride vapor will condense to a few drops of liquid, and air will re-enter the flask through the hole.

Remove the outer aluminum foil. Dry and weigh the flask, foil, and rubber band. Subtract the original weight from this value in order to determine the weight of the condensed liquid ethylene dichloride. Remove the inner foil and fill the flask with water to the very top. Pour the water into a graduated cylinder to find the volume of the flask.

You now have all the information you need! The weight of the liquid ethylene dichloride divided by the volume of the flask will be the density of the gas at $373°K$

and 1.0 atm pressure. (Small variations in external pressure can be neglected, but at high altitudes the actual barometric pressure in atm should be used.) Use Equation 10-8 to determine the molecular weight of ethylene dichloride. The molecular formula for ethylene dichloride is $C_2H_4Cl_2$. What is its true molecular weight? How close is your result?

Make a list of sources of error in this experiment and state whether each error would tend to increase or decrease your measured molecular weight.

### 2.   Cleaving Crystals

When an ionic crystal is shattered it tends to come apart, or cleave, to form pieces with plane faces intersecting at definite angles. This reflects the fact that such crystals have a highly regular internal structure. (See Fig. 10-19.) This contrasts with the random shattering of such non-crystalline solids as glass.

Carefully examine a 1 cm $\times$ 1 cm $\times$ 0.2 cm crystal of sodium chloride (such crystals can be obtained from the Harshaw Chemical Company of Cleveland, Ohio). Place a single edged razor blade on the large top surface of the crystal and hold it in a position parallel to one of the small faces. Tap the blade with a hammer. The crystal will cleave into two pieces. Take one of the two pieces and cleave it by the same technique in a direction perpendicular to the first cleavage. Take the other piece and attempt to cleave it at an arbitrary angle. What do you observe? Take one of the small cleaved pieces and tap it lightly with the hammer until it shatters. Examine the fragments with a magnifying glass. Using the remaining small pieces of crystal, see if you can cleave any of them in a direction not parallel to one of its faces.

An ionic crystal will generally have several planes along which it will cleave. These correspond to planes of atoms in the internal structure. The cleavage planes will be those that contain both positive and negative ions rather than those that contain either all positive or all negative ions. Why do you suppose this is true?

### 3.   Lots of Gas from a Little Bit of Solid

**CAUTION!** Dry ice can cause freeze burns if it comes in contact with the skin. Use tongs or asbestos gloves to handle the dry ice.

As we have noted, the transition of a solid to a liquid involves a small change in volume which can be either an increase or, for a smaller number of substances, a decrease. When either of these condensed phases becomes a gas a much larger change takes place. A convenient material with which we can measure the magnitude of this increase in volume is solid carbon dioxide (dry ice), which changes directly to a gas at room temperature and atmospheric pressure.

Obtain a small piece of dry ice. Using a millimeter ruler, estimate its volume. If it is larger than 0.8 cm$^3$ cut off a piece and re-estimate the volume. Place this small piece of dry ice in a 1000 ml graduated cylinder which has been filled with water and inverted in a large pan or tray containing water to a depth of a few cm. The dry ice will vaporize and displace water from the cylinder. When the solid has completely vaporized, measure the volume of the gas that was produced. (Some carbon dioxide will dissolve in the water but not enough to seriously affect your result.)

What is the approximate volume ratio of equal masses of gaseous and solid carbon dioxide? This ratio is typical of that which would be obtained with other materials.

problems

1    What is the numerical value for the universal gas constant if the volume is expressed in $cm^3$, the pressure in torrs, and the temperature in kelvins?

2    Suppose you had a closed container of gas initially at room temperature (21°C) and at a pressure of 760 torr and you heated it in a fire to 800°C. What will be the final pressure in the container? (Assume that the container doesn't expand when heated.)

3    How many liters of carbon monoxide (CO) gas will combine with 6 liters of oxygen ($O_2$) gas to form carbon dioxide ($CO_2$) gas? If the gases are at 0.9 atm and 20°C how many grams of $CO_2$ will be produced?

4    Suppose you had 1.14 grams of octane ($C_8H_{18}$) gas plus 4.64 grams of oxygen ($O_2$) gas confined in a 3.0 liter cylinder and you ignited the mixture with a spark, causing the following reaction to occur:

$$2\,C_8H_{18} + 25\,O_2 = 16\,CO_2 + 18\,H_2O$$

If the final temperature of the gas mixture produced is 600°C, what will be the final pressure in the cylinder? (In doing this problem it is important to note that only one of the starting gases will be completely used up.)

5    An important component of natural gas is a gaseous compound which has a density of 0.67 g/liter at a pressure of 1.0 atm. and a temperature of 20°C. What is the molecular weight of this substance? If 25 percent of its mass is hydrogen, how many hydrogen atoms are there in each of its molecules? Do you know the name of this substance?

6    A common demonstration experiment designed to illustrate Graham's Law of gas diffusion involves the simultaneous introduction of ammonia ($NH_3$) and hydrogen chloride (HCl) gases into opposite ends of a long horizontal glass tube. The gases diffuse down the tube and when they meet, a white ring forms because of the formation of the solid substance ammonium chloride ($NH_4Cl$). What will be the ratio of the distance of this white ring from the end of the tube where $NH_3$ was introduced to its distance from the end where HCl was introduced?

7    Why do you suppose $\frac{3}{2}k$ was chosen as the proportionality constant between molecular kinetic energy and absolute temperature (Equation 10-19) instead of simply $k$?

8    Suppose the volume of the gas sample that was used in deriving the gas law from kinetic theory is reduced by transferring the gas to a smaller cubical container at the same temperature. This reduction in volume will result in an increase in pressure. Why will more pressure be exerted on the walls of the container even though the kinetic energy of the molecules will be the same? (In answering this question refer to the steps in the derivation affected by reducing the container volume.)

9    Describe the properties that distinguish a liquid from a gas, using operational language.

10    Which of the following two-dimensional figures can be used to make a space-filling two-dimensional pattern: squares, circles, equilateral triangles, regular pentagons, regular hexagons? Can any regular figure with more than six sides be used?

11    Annealing involves heating a solid to a temperature just below its melting point and keeping it at that temperature for several hours (or sometimes longer) before allowing it to cool slowly. Crystals which have been formed by rapid cooling of a liquid usually become harder and stronger after annealing. Why do you think this is true?

12    A national weather report for a day in July might list the following temperatures and relative humidities:

| City | Temp. (°F) | Rel. Hum. (%) |
| --- | --- | --- |
| Phoenix, Ariz. | 104 | 11 |
| San Francisco, Calif. | 72 | 52 |
| Houston, Tex. | 93 | 82 |
| New York, N.Y. | 88 | 96 |

What was the partial pressure of water in the air in each city when the measurement was taken? (To answer this you will have to consult a reference book that lists the vapor pressures of water at various temperatures.) If you condensed all of the water in the 1.0 liter air sample taken in each city, how many grams of water would you obtain in each case?

13    Look up the boiling points of HF, HCl, HBr, $NH_3$, $PH_3$, $CH_4$, and $SiH_4$. Compare them to the boiling points of $H_2O$ and $H_2S$. Which of these molecules undergo extensive hydrogen bonding in the liquid state?

# writing numbers in power-of-ten notation

Here's an example of a very simple calculation. The problem is to find the time taken to travel a particular distance if the velocity is known. Since $v = \Delta x / \Delta t$, the ratio of distance traveled to the time taken, the time taken must be: $\Delta t = \Delta x / v$. We find the time by a simple division of distance by velocity.

Now we'll use this expression to find the time taken for a light signal traveling at three hundred million meters per second to go past a proton with a diameter of one millionth of a billionth of a meter. The time is:

$$\Delta t = \frac{0.000{,}000{,}000{,}000{,}001 \text{ meter}}{300{,}000{,}000 \text{ meters/second}}$$

$$= 0.000{,}000{,}000{,}000{,}000{,}000{,}000{,}003 \text{ seconds.}$$

Clearly, we can't go on like this! What this subject needs is a better notation. Fortunately it exists, although unfortunately it is sometimes known in the schools as "the scientific notation". Since even lawyers make use of the method, we perfer to call it the "power-of-ten" notation.

The method is, first of all, just a way of keeping track of zeroes and the position of the decimal point. Here are some fundamental identities:

| $10^{-3}$ | $10^{-2}$ | $10^{-1}$ | $10^{0}$ | $10^{1}$ | $10^{2}$ | $10^{3}$ | $10^{4}$ | $10^{5}$ | $10^{6}$ |
|---|---|---|---|---|---|---|---|---|---|
| ↓ | ↓ | ↓ | ↓ | ↓ | ↓ | ↓ | ↓ | ↓ | ↓ |
| 0.001 | 0.01 | 0.1 | 1 | 10 | 100 | 1000 | 10,000 | 100,000 | 1,000,000 |

Notice the sensible meaning of the positive powers. The square of 10, $10^2$, is equal to 100. Similarly, the cube of 10 is equal to 1000. Some of the numbers have common names: $10^2$ is a hundred; $10^3$ is a thousand; $10^6$ is a million; $10^9$ is a billion.

The negative powers represent reciprocals of the number. For instance, $10^2$ is 100, and $10^{-2}$ is 1/100, or 0.01.

Another way of remembering the significance of the exponents, is that they indicate the location of the decimal point. After writing down the initial number (such as 1), move the decimal point the number of places given by the exponent; to the right for plus, to the left for minus. For instance, for

$$0123456$$
$$\downarrow\downarrow\downarrow\downarrow\downarrow\downarrow\downarrow$$

$10^4$, write 10000000 and place the decimal point 4 places to the right. That yields 10,000.0. For $10^1$, the decimal is one place to the right of the 1: $10^1 = 10.0$. For $10^0$, the decimal is zero places removed from the 1: $10^0 = 1.0$. For $10^{-1}$, the decimal is one place from the one, but to the left: $10^{-1} = 0.1$. For $10^{-3}$, the decimal is three places to the left of the 1: $10^{-3} = 0.001$.

So far this system might save some writing space, but otherwise seems unnecessary. Now, however, we'll see the real power of the method. Let's multiply two numbers together:

$$10^2 \times 10^3 = 100 \times 1000 = 100{,}000 = 10^5$$

Notice that *multiplication* can be done by *adding* the zeroes, or by *adding* the exponents. The rule is general and holds for both positive and negative exponents. Here are some examples:

1   $10^1 \times 10^4 = 10 \times 10{,}000 = 100{,}000 = 10^5$   $(1 + 4 = 5)$
2   $10^{-2} \times 10^2 = 0.01 \times 100 = 1 = 10^0$   $(-2 + 2 = 0)$
3   $10^{-1} \times 10^3 = 0.1 \times 1000 = 100 = 10^2$   $(-1 + 3 = 2)$
4   $10^{-4} \times 10^2 = 0.0001 \times 100 = 0.01 = 10^{-2}$   $(-4 + 2 = -2)$

Incidentally, notice that Example 2 makes it apparent why $10^0 = 1$. This is the result that is obtained by multiplying $10^{-1} \times 10^1$, or $10^{-3} \times 10^3$, or any similar positive and negative pair.

*Division* of numbers in this notation consists of *subtracting* the exponents. Here are some examples:

1   $\dfrac{10^3}{10^1} = \dfrac{1000}{10} = 100 = 10^2$   $(3 - 1 = 2)$

2   $\dfrac{10^6}{10^3} = \dfrac{1{,}000{,}000}{1{,}000} = 1{,}000 = 10^3$   $(6 - 3 = 3)$

3   $\dfrac{10^{-1}}{10^2} = \dfrac{0.1}{100} = 0.001 = 10^{-3}$   $(-1 - 2 = -3)$

4   $\dfrac{10^2}{10^{-1}} = \dfrac{100}{0.1} = 1000 = 10^3$     $(2 - (-1) = 3)$

5   $\dfrac{10^{-3}}{10^{-4}} = \dfrac{0.001}{0.0001} = 10 = 10^1$     $(-3 - (-4) = 1)$

If the "power-of-ten" notation simply saved writing out zeroes, and allowed us to do simple multiplication and division of decimal multiples, it would be useful only occasionally. Here's an even more powerful use of the notation: When we write $10^2$, we mean $1 \times 10^2$, or 100. We can also use the system to define $2 \times 10^2$, which would be 200. Indeed, any number can be written in terms of one or more digits times some power of 10. Here are some examples:

1   $30 = 3 \times 10^1$
2   $38 = 3.8 \times 10^1$
3   $6000 = 6 \times 10^3$
4   $608.3 = 6.083 \times 10^2$
5   $0.83 = 8.3 \times 10^{-1}$
6   $0.0431 = 4.31 \times 10^{-2}$
7   $300{,}000{,}000 = 3 \times 10^8$

Multiplication and division of any numbers can now be done in this notation.

Express all numbers in a problem in terms of a number with one digit to the left of the decimal, times a power of 10. Then multiply and divide the digits while adding or subtracting the powers of 10.

In the case of our first problem, we could find the time taken for light to pass a proton, by writing:

$$\dfrac{1 \times 10^{-15}\text{ m}}{3 \times 10^8\text{ m/sec.}} = \dfrac{1}{3} \times 10^{-23}\text{ sec.} = 0.3 \times 10^{-23}\text{ sec.} = 3. \times 10^{-24}\text{ sec.}$$

Notice that in the final answer, the number was transformed so that there would be one digit to the left of the decimal. Since that increased the number by 10, the exponent had to be reduced by a factor of 10 from $10^{-23}$ to $10^{-24}$. There is no law that requires each number to be written with just one digit to the left of the decimal. It's a convention, and sometimes it's more convenient to have some other combination.

Here are some examples of power-of-ten computations:

1   $\dfrac{684}{21} = \dfrac{6.84 \times 10^2}{2.1 \times 10^1} = \dfrac{6.84}{2.1} \times 10^1 = 3.3 \times 10^1 = 33$

2   $\dfrac{0.0084}{42000} = \dfrac{8.4 \times 10^{-3}}{4.2 \times 10^4} = 2.0 \times 10^{-7}$

$$3 \quad \frac{(0.45)(8352)}{(73)(0.0006)} = \frac{(4.5 \times 10^{-1})(8.352 \times 10^3)}{(7.3 \times 10^1)(6 \times 10^{-4})} = \frac{(4.5)(8.4) \times 10^2}{(7.3)(6) \times 10^{-3}}$$

$$\approx \frac{37}{45} \times 10^5 \approx 0.8 \times 10^5 = 8 \times 10^4$$

In the middle of example (3), notice that 8352 was rounded off to $8.4 \times 10^3$. Since one of the numbers in the product (0.0006) has only one significant figure (the 6), there's no point of carrying more than two during the calculation. Only one significant figure will be justified in the final answer. At the point where the digits and powers of 10 were separated, an eyeball approximation was made to get an approximate value. Usually at that point you should also do the actual calculation, perhaps with a slide rule. The final answer, given to only one significant figure could have been left as $0.8 \times 10^5$. Alternatively, the digit was moved to the left of the decimal. Since that increased the number by ten, the exponent had to be lowered by a factor of 10.

For most of the calculations in this book, as well as for most of the calculations of daily life, one or two significant figures are all that are justified or necessary. Calculations then become easy if the numbers are written in power-of-ten notation. Simply set up the arithmetic in a way similar to that used in problem (3). Add and subtract the powers of 10, and do a mental calculation of the digits to one significant figure. That answer may be good enough. If another significant figure is required, use a slide rule or do the arithmetic long hand. Your mental calculation at least gives you a figure that must be close to any computed answer. If not, check the computation for arithmetic mistakes or misplacement of decimal point.

# fussiness, precision, and common sense

People often think that physics is the science of precision, which usually means fussiness. Onward, ever onward, to the next decimal place! Occasionally, for very good and special reasons, physicists and chemists do go to great trouble to make extremely precise measurements. For many purposes, however, there is no need for precision, and therefore it should not be obtained or specified. If anything, physics is the science of common sense. What is required is to use only the precision necessary. Not less, and equally important, *not more*.

**Order-of-Magnitude Calculations.** In this book you will make frequent use of order-of-magnitude calculations. You will be asked to determine whether two quantities are *about* equal, or whether some number is closer to ten, or one thousand, or one million. Such practices are not suggested in order to save time or work. The order-of-magnitude approach is the common working mode of research scientists. Because it is so different from the way calculations are usually done in the schools, this free-wheeling method takes some getting used to. Developing skill in the method is worth the trouble, however. Order-of-magnitude calculations allow you to zero in on a problem, making quantitative what otherwise would be vague, qualitative, and perhaps even wrong. Here's an example for consideration.

Will a lake serving as a town water reservoir be polluted if one kilogram of LSD is thrown into the water? It takes only 100 micrograms ($10^{-4}$ g) of LSD to affect a person. The amount released, if evenly distributed, could affect ten million people. ($10^3$ g$/10^{-4}$ g $= 10^7$). The situation sounds grim, but we don't really know unless we use a few numbers. Does the stuff mix evenly in the water? If so, how big is the lake? Let's assume that it is roughly one kilometer by one-half kilometer and has an average depth of 10 meters. The volume is, therefore:

$$(1 \times 10^3 \text{ m})(5 \times 10^2 \text{ m})(1 \times 10^1 \text{ m}) = 5 \times 10^6 \text{ m}^3.$$

If the LSD is evenly mixed, the amount per cubic meter is $1 \times 10^3$ g/5 × $10^6$ m³, which equals $2 \times 10^{-4}$ g/m³. A cubic meter contains $10^3$ liters, and each liter is about one quart. The amount of LSD per liter is $2 \times 10^{-7}$ g/1. In order to get the threshold dose of 100 micrograms ($10^{-4}$ g), a person would have to drink 500 liters of water: ($10^{-4}$ g/2 × $10^{-7}$ g/1).

This order-of-magnitude calculation by itself does not resolve the issue of safety. There are too many unknown factors, particularly the question of the extent of mixing. However, it does set the whole problem in perspective. Frequently such calculations show that something definitely will, or will not, occur, and no further calculation is necessary. Even when the answer is borderline, the calculation usually provides some indication about which data or factors must be better known for the next approximation. Certainly, in the example that we just worked out, there would be no point in making detailed calculations of the area and depth of the lake in order to find its volume. That figure need not be known to better than a factor of two.

**Significant Figures.**   There is a standard convention for noting the precision with which a number is known. It is the set of rules concerning "significant figures." Sometimes the rules get more elaborate than is justified, but the main procedures should be observed in all calculations.

The significance of writing down a particular digit in a number is that we know the value is closer to that digit than to the one higher or lower. For instance, if we write the number 43, we are asserting that the actual value is somewhere between 42.5 and 43.5 (Note that there is no need to quibble about whether the upper limit should be 43.4 or 43.5). Here are some examples of significant figures of various numbers. Note that in each case the number is also given in power-of-ten notation. Sometimes it is easier to see the significance of the figures that way.

| | | |
|---|---|---|
| 43 | $4.3 \times 10^1$ | 2 sig. fig. |
| 4300 | $4.3 \times 10^3$ | 2 sig. fig.° |
| 0.0043 | $4.3 \times 10^{-3}$ | 2 sig. fig. |
| 4301 | $4.301 \times 10^3$ | 4 sig. fig. |
| 0.004301 | $4.301 \times 10^{-3}$ | 4 sig. fig. |
| 1,000,000 | $1 \times 10^6$ | 1 sig. fig.° |

If you say that a number has the value of 5, you are saying that you know the value is between 4.5 and 5.5. Of course, if the number refers to things that come in units—like people—then 5 means exactly 5. On the other hand,

---

° When these numbers are written as shown in the left hand column, there is ambiguity about the significance of the final zeros. Use of the power-of-ten notation avoids this problem.

if you say that 1000 people are in an auditorium, you probably mean there are between 950 and 1050 people. Or perhaps you can only judge to plus or minus 20%, in which case there are between 800 and 1200 people. As you can see, the use of significant figures can be ambiguous.

In addition or subtraction, the sum or difference has significant figures only in the decimal places where *both* of the original numbers have significant figures. Note that this does not mean that the sum cannot have more significant figures than one of the original numbers. In example 4 below, the number 0.003 has only one significant figure, but the sum has four. It is the decimal *place* of the significant figure that is important in addition and subtraction. In examples 5 and 6 there is another example of the ambiguity of final zeros in numbers. If you estimate that there are 400 people in a hall, meaning that there are between 350 and 450, then your estimate is not changed if 3 people leave. On the other hand, if you have $400 in a bank, and draw out $3, you have $397 left.

1    6.8        2    6.843        3    6.843
   $+1.1$          $+1.1$            $+1$
   ─────          ─────            ────
    7.9            7.9              8

4    6.843       5   400         6   $4.00 \times 10^2$
   $+0.003$         $-3$             $-.03 \times 10^2$
   ──────          ────             ──────────────
    6.846           400             $3.97 \times 10^2$

Bet you didn't know that rock is one million and ten years old.

How on earth do you know that ?

Ten years ago another scientist fellow was lookin' at that rock and he told me then that it was a million years old.

1,000,000 + 10 = 1,000,000

In multiplication or division, the product or quotient cannot have more significant figures than there are in the least precisely known of the original numbers. If you multiply a number with two significant figures by a number with three significant figures, the product can have only two significant figures. Here are some examples.

$$\begin{array}{r} 5.2 \\ \times 3.1 \\ \hline 52 \\ 156 \\ \hline 16.12 = 16 \end{array}$$

$$5.243 \times 3.1 = 16$$

$$5.243 \times 0.0031 = 0.016$$

$$\frac{37}{9} = 4$$

$$\frac{37}{9.1} = 4.1$$

Usually, during multiplication or division, an extra significant figure is carried along, and the final answer is then rounded off appropriately.

**Functional Dependence.**    In many cases, in everyday life as well as in science, we look for relationships between variables without regard to numerical quantities. If you double your car speed, do you reduce your travel time by a factor of two? Such a question concerns functional dependence. If two variables, $x$ and $y$, are related, does $x = Ky$, or does $x = K/y$, or does $x = Ky^2$? Frequently this sort of information is more useful than any details of actual magnitudes of the numbers.

**Precision of Speech.**    Every field of human endeavor develops its own peculiar language and customs concerning the use of special words. Unfortunately, the language and customs usually consist of subtle and nice variations on the standard vernacular. Sometimes the special definitions are justified because they reflect profound relationships. Occasionally the jargon is petty and serves to bar the outsider to the profession or becomes the subject of test items for the student. Worse yet, in many cases members of a profession will have their own particular verbal fetishes and will not agree with their colleagues about whether the proper use of a particular word is silly or vital. Here are some examples:

Technically, at least in the schools, *velocity* is a vector, which means that it signifies both magnitude and direction (as well as having certain combinatorial properties). Speed, on the other hand, is a scalar, signifying only the magnitude of a velocity. In this text, as in most of the real world, we use these words interchangeably. The context invariably provides sufficient information about the nature of the quantity. Besides, $v$ is the only convenient symbol for velocity; $s$ often means displacement.

Heat is internal energy in transit between two bodies, often (but not always) caused by a temperature difference. If you "heat up" an object you do not necessarily raise its temperature. You might, for instance, only melt it while it remains at a constant temperature. Once heat enters an object, it does not reside there as heat. The phenomenon is better understood by referring to the stored energy as *internal* energy. Some of the internal energy is in the form of potential energy of distortion and some in the vibrational energy of the atoms. Here is a case where precision of language is needed in order to clarify basic concepts.

The field of electricity is rife with special word uses. A battery, for instance, is a connected group of electrical cells. No great issue is at stake here, however. Like the nearest hardware salesman, we frequently use the word battery to mean cell. On the other hand, electrical currents really shouldn't *flow* in wires. Charges flow; currents exist. Such a distinction is a minor point, one to be observed in writing a text, but not in correcting an exam.

It is not possible to set hard and fast rules in this semantic game. The boundaries between fussiness and precision continually shift. Frequently there is more concern about niceties of definitions among those in academic life than among those in active research. In lieu of rules, a good science course should offer examples and practice—the standard situation for any art.

# index